teach yourself®

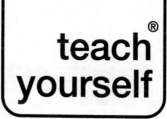

planets
david a. rothery

For over 60 years, more than
50 million people have learnt over
750 subjects the **teach yourself**
way, with impressive results.

be where you want to be
with **teach yourself**

For UK order enquiries: please contact Bookpoint Ltd, 130 Milton Park, Abingdon, Oxon, OX14 4SB. Telephone: +44 (0) 1235 827720. Fax: +44 (0) 1235 400454. Lines are open 09.00–17.00, Monday to Saturday, with a 24-hour message answering service. Details about our titles and how to order are available at www.teachyourself.co.uk

For USA order enquiries: please contact McGraw-Hill Customer Services, PO Box 545, Blacklick, OH 43004-0545, USA. Telephone: 1-800-722-4726. Fax: 1-614-755-5645.

For Canada order enquiries: please contact McGraw-Hill Ryerson Ltd, 300 Water St, Whitby, Ontario, L1N 9B6, Canada. Telephone: 905 430 5000. Fax: 905 430 5020.

Long renowned as the authoritative source for self-guided learning – with more than 50 million copies sold worldwide – the **teach yourself** series includes over 500 titles in the fields of languages, crafts, hobbies, business, computing and education.

British Library Cataloguing in Publication Data: a catalogue record for this title is available from the British Library.

Library of Congress Catalog Card Number: on file.

First published in UK 2000 by Hodder Education, 338 Euston Road, London, NW1 3BH.

First published in US 2000 by The McGraw-Hill Companies, Inc.

This edition published 2003.

The **teach yourself** name is a registered trade mark of Hodder Headline.

Copyright © 2000, 2003 David A. Rothery

In UK: All rights reserved. Apart from any permitted use under UK copyright law, no part of this publication may be reproduced or transmitted in any form or by any means, electronic or mechanical, including photocopy, recording, or any information, storage and retrieval system, without permission in writing from the publisher or under licence from the Copyright Licensing Agency Limited. Further details of such licences (for reprographic reproduction) may be obtained from the Copyright Licensing Agency Limited, of Saffron House, 6–10 Kirby Street, London, EC1N 8TS.

In US: All rights reserved. Except as permitted under the United States Copyright Act of 1976, no part of this publication may be reproduced or distributed in any form or by any means, or stored in a database or retrieval system, without the prior written permission of the publisher.

Typeset by Transet Limited, Coventry, England.
Printed in Great Britain for Hodder Education, a division of Hodder Headline, 338 Euston Road, London, NW1 3BH, by Cox & Wyman Ltd, Reading, Berkshire.

The publisher has used its best endeavours to ensure that the URLs for external websites referred to in this book are correct and active at the time of going to press. However, the publisher and the author have no responsibility for the websites and can make no guarantee that a site will remain live or that the content will remain relevant, decent or appropriate.

Hodder Headline's policy is to use papers that are natural, renewable and recyclable products and made from wood grown in sustainable forests. The logging and manufacturing processes are expected to conform to the environmental regulations of the country of origin.

Impression number   10 9 8 7 6 5
Year   2010 2009 2008 2007

contents

acknowledgements

After the success of *Teach Yourself Geology*, it seemed natural to follow it up with a book such as this, which I hope will be appreciated by a similar but even wider readership. I am extremely grateful to Alex Jolesz for commenting on the first draft so promptly, and for the trouble she took to help me clarify what I was trying to say. If any sentences are still too long, it's not her fault. It's mine. During the three years since I wrote the first edition of this book, the water on Mars story has continued to unfold, thanks largely to detailed images from orbiting spacecraft. Advanced telescope work has led to the discovery of dozens more tiny moons of Jupiter and Saturn, and changed our understanding of minor planetary bodies both in near-Earth space and on the Solar System's outskirts beyond Neptune. You will find these and other recent developments reflected in this new edition.

Line drawings in this book are my own. Unless credited alternatively below, the 'photographic' images derive from NASA (National Aeronautics and Space Administration). Without NASA's exploration of the Solar System (funded by the taxpayers of the USA) this book could scarcely have been written. I acknowledge gratefully the following additional sources of images: Figure 5.2 National Oceanic and Atmospheric Administration; Figures 7.6 (right), 7.7, 7.8 (right), and 7.9 NASA/JPL/Malin Space Science Systems; Figures 7.2, 9.7 (lower left and right), 10.1, 11.2 and 13.1 (top and middle left) and Plate 10 NASA and AURA STScI; Figure 10.13 ESA/NASA/JPL/University of Arizona; Figure 13.1 (bottom) courtesy of Marc W Buie/Lowell Observatory; Figure 13.2 NASA and C. Veillet (Canada-France-Hawaii Telescope).

01

introduction

In this chapter you will learn:
• how this book is organized, and about the cultural and historical significance of our Sun's planets.

There can be few of us who, happening to glance at the western sky after sunset, have not noticed with pleasure that the bright white beacon of Venus has returned to the evening sky. Equally, most of us have probably been fooled sometime by the sight of an aircraft's landing light, mistaking it for a bright planet such as Venus or Jupiter. We may not realize our deception unless we notice the aircraft is moving relative to the background stars.

As with other objects in the night sky, the planets are now much less apparent to the casual observer than they were before the days of bright streetlighting, floodlit sports fields, and air traffic. They *can* still be seen and recognized if you know what to look for, and I encourage you to try. Merely spotting them in the sky is satisfying, and seeing with your own eyes through a telescope the shimmering globe of another planet provides a frisson that never quite goes away. As for our own Moon, it is so close that even binoculars can reveal gorgeous views of its varied landscapes.

However, for me, the main appeal of the planets is that their surfaces, and those of their larger satellites, have been revealed in stunning detail by spaceprobes, often exceeding in detail what you could see on the Moon through a large telescope. There are now dozens of worlds whose surface and atmospheric composition have been determined, and whose landscapes and weather patterns are known in sufficient detail for us to get a real taste of what it would be like to be there.

My main purpose in this book is to describe these worlds, and in doing so to reveal the processes that have shaped them and which in many cases continue to do so today. I therefore deal largely with phenomena that could be considered geological (volcanic eruptions, fault movements, erosion, and so on) and meteorological (such as climate change and atmospheric circulation) rather than astronomical. I make no apology for this, because most of the researchers who study planetary bodies today refer to themselves as planetary scientists rather than astronomers. They are more likely to be qualified in geology, geophysics or atmospheric science than in any branch of astronomy.

After giving a brief overview of the Solar System and its origin, I describe each planet in turn. The Moon and the asteroids get separate chapters. In the chapters on Jupiter, Saturn, Uranus and Neptune you will find just as much about their major satellites. These are large and fascinating bodies in their own right, and any description of planetary bodies in our Solar System would be woefully incomplete without them. To illustrate my account,

I use as many close-up images as space allows. I have avoided jargon where possible, but some technical terms are unavoidable. These are shown in bold typeface where they make their first major appearance. I have tried to make the meaning clear in each case, but have thought it advisable also to provide a list of definitions in the glossary.

I hope that by the time you reach the end of the book you will want to find out more about the planets. If so you should find the 'taking it further' section useful. It has sufficient hints on observing the planets to get you started, and a list of websites where you can inspect for yourself the wealth of images and other information now available on the planets and their satellites. As a prelude though, first a little about the cultural and historical significance of the planets.

History of the planets

There were five **planets** (other than the Earth) known in the ancient world. These are the ones bright enough to be noticeable to the unaided eye, and are the planets we now know as Mercury, Venus, Mars, Jupiter and Saturn. They all appear bright, and all except Saturn can outshine the brightest star, though Mercury is hard to spot because it is always fairly close to the Sun in the sky. However, what made the planets special to our remote ancestors was not their brightness but the fact that they change their positions whereas the stars remain in fixed patterns. There is nothing spectacular in this motion; unless you are using a telescope you need to be patient, and note the position of a planet relative to background stars on successive nights in order to perceive that it is indeed slowly wandering about the sky. The term 'planet' in fact comes from an ancient Greek word for wanderer. The names of the individual planets, which they have borne in the western world from antiquity, are those of the gods of the classical Greek and Roman world, though other cultures such as Chinese and Babylonian identified these planets as special objects too.

The Sun and the Moon are much brighter than the planets and show perceptible discs. The Sun is much further away from us than the Moon, and the two appear virtually the same size only because the Sun is proportionately much larger than the Moon. The ancient Greeks realized this, and correctly regarded the Moon as the closest celestial object to the Earth.

Most classical cosmological theories had everything revolving about the Earth. It was not until the seventeenth century (several decades after Nikolas Copernicus published his famous heliocentric, or Sun-centred, theory in 1543) that it became widely accepted that the Moon is the *only* body to orbit the Earth, and that the Earth and the other planets all go round the Sun.

We now realize that the Earth is just another planet among many. Right from their invention, telescopes played a leading role in increasing our understanding in this respect. Using one of the first telescopes in 1610, Galileo Galilei not only saw mountains on the Moon and documented the phases of Venus, but he also discovered four **satellites** in orbit around Jupiter, proving that the Earth is not the centre of all motion. By 1700 five of Saturn's satellites had been observed. In 1781 William Herschel discovered the planet Uranus, and within six years he had found two of its satellites. Neptune, whose location had been predicted from its perturbation on the orbit of Uranus, was first seen in 1846 and Pluto was discovered in 1930.

Astrologers claim (without any foundation) that the positions of the planets in the sky at the hour of your birth determine both your character and, to an extent, your fate. Each newly discovered planet (though not, as far as I know, their satellites) was incorporated into the astrological scheme, silently ignoring the implication that former horoscopes, which were based on incomplete data, must have been badly flawed if not downright wrong. The more conscientious astrologers are now struggling to come to terms with the recent realization that Pluto, although for historical reasons still classified as a planet by the International Astronomical Union, is merely the largest known member of a vast horde of icy bodies orbiting the Sun in Trans-Neptunian space! The rational, scientific, view is that the planets are too far away for their gravitational attraction (the only known force that can extend over interplanetary distances) to have any perceptible short-term effect upon conditions on the Earth or its inhabitants' lives.

The true significance of the other planets lies not in the charade of astrology, but in what they have taught us about our home planet. Paradoxically, this is both that the Earth is nothing special and that it is also very special indeed. On the one hand, the Earth is unremarkable in that it is a planetary body much like any other (though each has its own special characteristics). On the other hand, the Earth is very special in that it is our home and the only place in the Solar System that appears to be capable of supporting abundant and complex life. However, as we shall see, this does not mean that the Earth is necessarily the only place where life occurs.

02 the Solar System

In this chapter you will learn:
- the basic layout of the Solar System, what it is made of, and how bodies move round their orbits
- how modern theories can explain its origin
- what distinguishes a planet from other Solar System objects.

Local geography

The Earth and the eight other traditional planets go round the Sun. This is one of a hundred billion stars belonging to our galaxy, and is moving at a speed of about 220 km per second relative to the galactic centre, about which it is rotating. However, in a *local* geography lesson such as this we are concerned only with motion within the system of objects gravitationally bound to the Sun, in other words with the **Solar System**. In the description that follows, I describe motion relative to the Sun, as if the Sun occupied a fixed position in space. This is not merely a convenient fiction, it recognizes that the Sun is by far the dominant body in the Solar System. It has a thousand times the mass of Jupiter, which itself has more mass than all the other planets put together. The Sun's gravitational attraction therefore controls the motion of its attendant bodies; all the more so as the next nearest star is 40,000 billion km away. In contrast, the distance from the Sun to the Earth is nearly 150 million km, and the average distance of Pluto from the Sun is just under 6 billion km. To avoid using such (literally) astronomical figures, it is common to quote distances within the Solar System in terms of **Astronomical Units (AU)**, where 1 AU is the Earth–Sun distance.

It is not just distances that are enormous in the Solar System. The masses of planets are also too great to quote conveniently in everyday units. For example, the Earth has a mass of very nearly 6 million billion billion kg (6×10^{24} kg), and it is often convenient to quote masses relative to this value. This has been done in Table 2.1 on the next page, which shows some basic planetary information.

Some pertinent facts emerge from study of this table. The first is that the Solar System is mostly empty space: the distances between planets are vastly greater than the sizes of the planets themselves. Another is that the further a planet is from the Sun, the longer it takes to complete an orbit of the Sun. This is not merely a result of the orbital track being longer with greater distance from the Sun, it is also because planets travel more slowly the further they are from the Sun. It was Johannes Kepler (1571–1630) who first worked out the correct relationships between distances and orbital speed, as stated in what are now referred to as **Kepler's laws of planetary motion**.

	Distance from Sun (AU)	Mass (relative to Earth)	Equatorial radius (km)	Orbital period	Rotation period
Mercury	0.39	0.055	2439	88.0 days	58.6 days
Venus	0.72	0.81	6052	224.7 days	243.0 days
Earth	1.0	1.0	6378	365.3 days	1.0 day
Moon	1.0	0.012	1738	(27.3 days)	27.3 days
Mars	1.52	0.11	3394	687 days	1.026 days
Jupiter	5.2	318	71,400	11.86 years	0.410 days
Saturn	9.5	95.2	60,000	29.46 years	0.444 days
Uranus	19.1	14.5	25,600	84.01 years	0.718 days
Neptune	30.0	17.2	24,300	164.8 years	0.768 days
Pluto	39.4	0.0022	1170	247.7 years	6.38 days

table 2.1 some basic data for the planets
time is quoted in Earth-days, and Earth-years
the orbital period given for the Moon is for its orbit about the Earth; together the
Moon and Earth orbit the Sun once per year (which is how a year is defined)

Kepler's laws

The relationship between average distance from the Sun and
orbital period is expressed in Kepler's third law of planetary
motion, which states that the square of a planet's period of
revolution round the Sun (its orbital period) is proportional to
the cube of its average distance from the Sun. However, it was
not until Isaac Newton (1642–1727) propounded his theory of
gravitation that the reasons were understood. The
understanding of gravity provided by Newton enables the mass
of the Sun to be calculated from the size and period of the orbit
of any of its planets. Similarly the mass of a planet can be
deduced from the orbital characteristics of any of its satellites.

The orbits of the planets are very nearly circles, but not quite,
and it was Kepler who first realized that they are ellipses. His
first law of planetary motion states that planets move in
elliptical orbits, with the Sun at one focus of the ellipse. An
ellipse can be thought of as an elongated circle; it has two foci,
positioned on the long axis of the ellipse such that the sum of
the distances between any point on the ellipse and the two foci
is constant (Figure 2.1). The more elongate the ellipse, the
further apart are its two foci and hence the greater the ratio
between the distance between the foci and the size of the ellipse's
long axis. This ratio is called the **eccentricity** of the ellipse.
Conversely, when the two foci coincide the eccentricity is zero

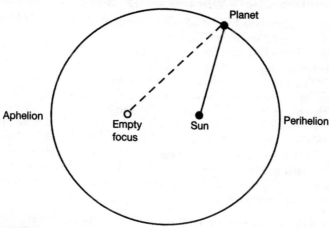

figure 2.1 an ellipse

the two dots (one black, one white) near the centre are the foci of the ellipse
these are shown joined to an arbitrary point on the ellipse by a solid line and
a dashed line respectively

for all such points on the ellipse, the lengths of the two lines add up to the
same amount

in the case of a planet orbiting the Sun, the planet would be represented by
the dot on the ellipse, and the Sun by the black dot at one of the foci

the other focus is empty

and the ellipse is simply a circle. The orbits of the major planets
have quite small eccentricities, 0.017 in the case of the Earth and
0.048 for Jupiter, which makes them pretty much
indistinguishable from circles when you see them drawn in a
diagram. The ellipse in Figure 2.1 has an eccentricity of 0.32,
which is greater than for any of the planets. However, it does
show clearly that the distance between a planet and the Sun
must vary continuously during the course of an orbit. The point
on the orbit closest to the Sun is called **perihelion**, and the
furthest point is called **aphelion** (from the ancient Greek for
'near the Sun' and 'far from the Sun', respectively).

Kepler's second law of planetary motion is a little harder to
visualize than the other two. It states that, for a given planet, a
line drawn from the planet to the Sun (the solid line in Figure 2.1)
will always sweep out the same amount of area over a given time
interval. The most important consequence of this is that a planet
travels fastest when it is at perihelion and slowest when it is at
aphelion.

Planetary orbits and planetary spin

Thus the planets orbit the Sun in near-circular ellipses, with the inner planets travelling much faster than the outer ones, and each one speeding up a little as its orbit takes it closer to the Sun and slowing down a little as its distance from the Sun increases again. To complete this basic picture, you need to know that the orbits of all the planets lie very nearly in the same plane and that all the planets orbit the Sun in the same direction, which is anticlockwise as seen from a point in space way above the Earth's (or the Sun's) north pole. This sense of anticlockwise motion is referred to as **prograde**, the opposite direction being called **retrograde**. In the case of satellites, orbital motion is defined as prograde if it is in the same direction as their planet's spin, which is usually the case.

The plane of the Earth's orbit is called the **ecliptic**, because the Moon can hide the Sun (i.e. cause an eclipse) only when it crosses this plane at exactly the same time as it passes between the Earth and the Sun. The Moon's orbit about the Earth is inclined at 5.2° to the ecliptic, and the orbits of the planets lie within 3.5° of the ecliptic except for Mercury (7.0°) and Pluto (17.2°). Because everything lies so close to this plane, a pretty good representation of the three-dimensional shape of the Solar System can be obtained by drawing it on a flat sheet of paper (Figure 2.2).

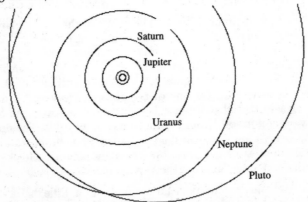

figure 2.2 the Solar System drawn to scale
the inner two rings mark the orbits of the Earth and Mars (the orbits of Venus and Mercury that lie within the orbit of the Earth being too small to include) also omitted are the orbits of the asteroids, of which there are several thousand (mostly between Mars and Jupiter), and hordes of small icy bodies in the Neptune-Pluto region
for scale, Neptune's orbit is 30 AU (4500 million km) from the Sun

The Sun itself and most planets rotate (i.e. spin on their axis of rotation) in the prograde direction, but no planet's axis of **rotation** is exactly perpendicular to its orbit. The amount (in degrees) by which a planet's rotation axis departs from being perpendicular to its orbit is referred to as its **axial inclination**.

This pattern of prograde rotation close to a common plane is shared by most smaller objects in the Solar System too, including such things (defined in subsequent sections) as the asteroids (concentrated between the orbits of Mars and Jupiter), icy Kuiper belt objects orbiting beyond Neptune, and short period comets. Only long period comets, hurtling inwards from 10,000 AU or more, seem immune and can come in from any angle and pass round the Sun in either direction.

The next section tells how the situation just described came about. In what follows, and throughout the rest of the book, you will find the handy term **planetary body** used to encompass both the true planets and other bodies (such as the major satellites) that are planet-like in many of their important properties.

A short history lesson

The planar shape and shared sense of rotation within the Solar System provide strong evidence for how the Solar System came into being. Our present understanding is that about 4.6 billion years ago the Sun formed by gravity-driven contraction of an interstellar gas cloud (mostly hydrogen) containing traces of dust (ices, carbon and specks of rock). The event can be dated by measuring the products of radioactive decay in meteorites.

Most of the matter in the contracting cloud ended up in the centre and formed the Sun. This began to shine at first because of the heat of all the infalling matter but pretty soon it was shining even more strongly because its centre became hot enough for hydrogen nuclei to fuse to form helium, which is what powers the Sun today. The original cloud was inevitably rotating slightly. As it contracted its rotation speeded up, because of 'conservation of angular momentum'. The same principle means that an ice skater who starts to spin with arms extended can make herself spin faster simply by drawing her arms inwards. Just as the hem of the ice skater's skirt would fly outwards as she span, so the cloud's rotation forced it into a disk shape. The planets grew within this disk, which is referred to as the **solar nebula** (Figure 2.3), and inherited their sense of rotation and orbital characteristics from it.

From dust to planetesimals

At first, virtually all the original dust in the solar nebula was vaporized by the heat of the young Sun, but as the nebula cooled down new dust grains began to condense. Because the nebula within a few AU of the Sun was dense (by the standards of outer space) collisions between particles were relatively common. Evidently many of the dust particles were fluffy, so that when they collided they tended to stick together. In this way, progressively larger particles were built up, taking maybe as little as 10,000 years to grow into globules a centimetre across. After about 100,000 years, random collisions had produced hordes of bodies about 10 km across, termed **planetesimals** ('tiny planets'). These were all swirling round the Sun in the

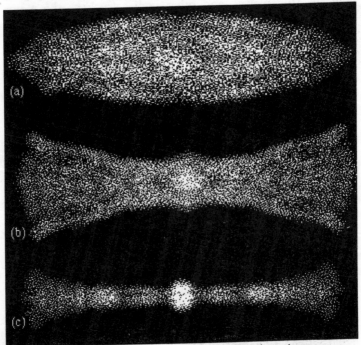

figure 2.3   cross-sectional view of the solar nebula at three stages, (a)–(c), covering a period of about 100,000 years while it was assuming a disk shape

the distance from centre to edge was about 40–50 AU (6000–7500 million km)

same prograde direction, and were now large enough for mutual gravitational attraction between planetesimals to begin to play a significant role, so collisions became more common. The largest planetesimals had the greatest pull, so these naturally suffered more collisions and grew fastest. Within a further few tens of thousands of years the largest planetesimals had grown to a few thousand km across, destroying most of the smaller ones in the process.

Planetary embryos and giant impacts

The largest planetesimals at this stage, of which there were perhaps a few hundred, are dignified by the name of **planetary embryos**. The era of frequent collisions between planetesimals was now over, because the spaces in between had become too great, and it must have taken maybe 50 million years to build up an Earth-sized planet by chance collisions between planetary embryos.

These collisions, known as **giant impacts**, would have been totally devastating. In some circumstances the two colliding bodies would have been fragmented by the force of the impact, but more often than not fragments of the smaller of the two must have become plastered over the surface of the larger. There was probably enough heat liberated during such a collision to melt the target body, which would have allowed denser materials to segregate inwards and lighter materials to work their way outward. In this way, the interior of each rocky body like the Earth (the **terrestrial planets**) was able to separate into an iron-rich **core** surrounded by a rocky **mantle**. This process is described as **differentiation**.

For many purposes a more important distinction within the outer zone of a planet than between crust and mantle, is between a strong-and-rigid outer shell and a soft-and-mobile interior. In the case of the Earth, its strong-and-rigid outer shell, known as the **lithosphere**, includes the crust and the topmost mantle in a single mechanical layer about 100 km thick. Below the lithosphere the mantle, though technically a solid, is able to flow and convects at a rate of a few centimetres per year. The convecting mantle is sometimes referred to as the **asthenosphere**.

The Earth's Moon apparently owes its origin to a giant impact in which much of the debris from the fragmented impactor plus some ejecta from the Earth ended up in orbit about the Earth (where it collected together to form the Moon) rather than back on its surface. Once the last giant impact in the inner Solar System had occurred, there were just four surviving terrestrial planets (Mercury, Venus, Earth and Mars). The total is five if you count the Moon, which is a planet in a geological sense because of its size and character, though not in an astronomical sense because it orbits the Earth rather than the Sun.

Impact cratering

The evolution and further internal differentiation of these bodies from then on was driven mainly from within, by processes that changed the chemical composition of the outermost part of the mantle sufficiently to merit its distinction by the term **crust**. However, the terrestrial planets were all subject to one very important external influence in the form of bombardment by 'junk' left over from the origin of the Solar System, including the surviving small planetesimals, which scarred their surfaces with **impact craters** (Figure 2.4).

Crater formation has continued up to the present day, but its intensity waned to something like its present level about 3.9 billion years ago, by when most of the original debris had been mopped up. Counting the density of impact craters on a planetary surface is virtually the only way we have to estimate the age of the surface (i.e. how long the material now forming the surface has been there). The basis of this **cratering timescale** is that older surfaces have had time to accumulate more craters per unit area than younger surfaces. Rock samples collected from the Moon and then dated in the laboratory (by means of their natural radioactivity) allow us to put numerical values to these ages, which would otherwise be a purely relative scale.

Growth of the giant planets

The processes and timescales described above seem to apply well in the inner Solar System, out as far as the orbit of Mars. Further from the Sun, planet formation happened slower and finished several million years later. The planets that grew there are much larger than those in the inner Solar System, and are known as the **giant planets**. Part of the reason for this is that at about 5 AU from the Sun, it was cold enough for frozen water (a very abundant substance, as you will see shortly) to condense directly from the solar nebula. The outer planets began as rock balls of several Earth-masses surrounded by ice. Once such a planet had reached about ten Earth-masses it became a very

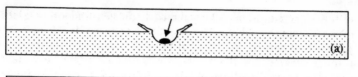

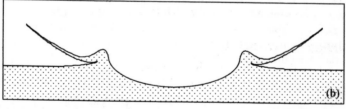

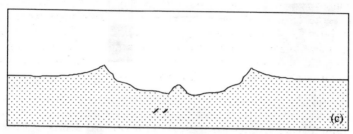

figure 2.4 cross-sections to show how an impact crater is formed, by collision at typical Solar System closing speeds of a few tens of km per second

(a) a high-velocity projectile strikes the surface

a shock wave begins to excavate a crater

it also destroys most of the projectile

(b) excavation by the shock wave proceeds radially out from the point of impact

as drawn at this stage the crater has virtually reached its final diameter (20–100 times that of the impactor) and the newly excavated material is barely able to flop over the rim

(c) a central peak has risen by rebound, and the inner walls of the crater have slumped, giving the crater its final form

where there is no atmosphere to shield the surface, impact cratering occurs at all scales from micrometres to a thousand kilometres

a single central peak, as shown here, occurs only across a limited size-range (15–140 km on the Moon, 7–70 km on the Earth)

it takes only about 100 seconds to excavate a 100 km diameter crater

material thrown out of an impact crater is described as **ejecta**, and may form a recognizable deposit on the surface

effective gravitational scavenger of hydrogen and other gases directly from the nebula, and so grew even more. In the cases of Jupiter and Saturn, the mass of hydrogen captured in this way is much greater than that of their rock and ice 'cores', and these two planets are sometimes described as the **gas giants**. The other two giant planets, Uranus and Neptune, are considerably less massive (Table 2.1) mainly because they were not able to capture so much gas. The internal structures of the giant planets today are compared in Figure 2.5.

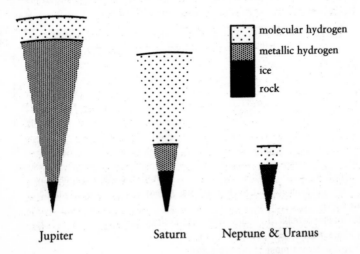

Jupiter   Saturn   Neptune & Uranus

figure 2.5   cross-sections to compare the internal structures of the giant planets
metallic hydrogen is hydrogen atoms packed so closely together that they behave like a metal
the hydrogen layers probably contain about 20 per cent of helium
(approximately the same hydrogen:helium ratio as in the Sun and the solar nebula)
ice is not necessarily just frozen water; it can include frozen methane, ammonia, carbon monoxide and nitrogen
for scale, Jupiter's radius is nearly 71,500 km

Uranus and Neptune differ from Jupiter and Saturn because, being further from the Sun, they grew more slowly. Their gas capture stage was prematurely ended when the remnants of the solar nebula were blown clear of the Solar System by an outflow of gas from the Sun, described as the 'T Tauri wind', after a young star called T Tauri that is going through this stage today.

This wind also swept away whatever primordial atmospheres had been scavenged from the solar nebula by the terrestrial planets, whose present atmospheres owe their origin largely to escape of gases from within (by volcanic activity and the like). The Sun's T Tauri phase lasted only about 10 million years. Since then, matter in the form of charged particles has continued to stream out from the Sun in a much more tenuous phenomenon described as the 'solar wind'. At 1 AU from the Sun, the speed of the solar wind is about 250 km per second, but its density is so slight that less than one-ten thousandth of the Sun's total mass has been lost in this way during the Sun's lifetime.

The giant planets had disturbing effects on their neighbour-hoods. Each was able to collect a disc of gas and dust around itself, perhaps resembling the solar nebula in miniature. Within these discs various bodies of small planetary size were able to grow and become the major satellites of the giant planets. Two of these satellites are bigger than Mercury, and there are several bigger than Pluto. These fascinating bodies are worlds in their own right, and are described individually in the same chapters as their planets. Each giant planet also has a system of rings, consisting of untold numbers of particulate fragments orbiting in the plane of the planet's equator and possibly representing the remains of a disrupted satellite. Even Saturn's extensive and highly reflective ring system contains no more mass than a single body 100 km across, and the other planets' rings are far less substantial even than this.

The asteroids

Jupiter is so massive that the effects of its gravitational field were felt beyond its immediate neighbourhood while the Solar System was taking shape. It stirred up the orbits of the planetesimals that were orbiting between Jupiter and Mars to such an extent that collisions between planetesimals were too violent to permit growth into larger bodies. Instead, collision often led to fragmentation rather than to growth by accretion, and so no planetary embryos were able to grow. Most of the material that formerly lay in this region of space has now been flung out of the Solar System as a result of close encounters with Jupiter. The bits that remain, amounting to less than a thousandth of an Earth-mass in total, are the **asteroids** (sometimes known as **minor planets**). The region of space between the orbits of Mars and Jupiter is sometimes referred to as the asteroid belt, though, as discussed in Chapter 08, the asteroids are by no means confined to that region.

Left-overs on the fringes

Beyond Neptune, the distances are so great that collisions between planetesimals were too rare to allow the formation of large planets. Between 1992 and 2002 several hundred icy bodies were discovered beyond the orbit of Neptune, mostly travelling in orbits between about 30 and 50 AU in a region known as the **Kuiper belt**. Estimates suggest that there are probably some 70,000 Kuiper belt objects more than 100 km in diameter, and untold hordes of smaller ones. It now seems that Pluto (discovered in 1930) is merely the largest of the independent Kuiper belt objects, and that Neptune captured its largest satellite (Triton, a body that is bigger than Pluto) from this belt.

Meteorites and comets

All classes of object described above are discussed in later chapters. It is a complete inventory of those Solar System objects that have characteristics qualifying them for consideration in a book about planets. There are two kinds of smaller bodies that will not be further discussed. These are meteorites and comets. **Meteorites** are lumps of rock ('stony meteorites'), or rock plus iron ('stony irons'), or nickel-iron ('iron meteorites') that fall to Earth and the vast majority of which are apparently debris from collisions between asteroids. Many show clear evidence that parts of their parent bodies were formerly molten. However, some stony meteorites appear to be relatively unaltered agglomerations of the material from which the Solar System formed. Notable among these is a group called **carbonaceous chondrites** that contain several per cent of carbon in the form of non-biologically produced organic molecules. Really small meteorites (less than a millimetre across) are called micrometeorites, and many of these may be dust that never formed part of a larger body, or that has escaped from comets.

Comets are mixtures of ice, carbonaceous material and rock dust a few km or tens of km across. They can develop long spectacular tails of gas and dust when their highly eccentric orbits bring them close enough to the Sun for the temperature to become high enough to vaporize the water, carbon monoxide and other volatile substances otherwise trapped as ice. Short-period and long-period comets have already been distinguished by their orbital characteristics. Short-period comets are probably small Kuiper-belt objects that have been gravitationally scattered inwards, and will survive only a few solar passages before losing all their volatiles. Long-period comets occasionally fall into the inner Solar System from a reservoir some forty thousand AU from the Sun known as the Oort cloud. These probably began life as small icy planetesimals in the Jupiter-Neptune region that were flung away from the Solar System by gravitational encounters with one of these giant planets.

A little bit of chemistry

So much, then, for the layout and history of the Solar System, but I have not said much about what it was made from, other than that the interstellar cloud from which the solar nebula formed was mostly gas with some dust. The abundances of the elements at the visible surface of the Sun can be determined quite accurately by splitting sunlight into a spectrum and measuring the strengths of absorption lines caused by the presence of each element. Because there is no known process capable of changing the composition of the outer part of the Sun through time, it is a pretty safe bet that this shows us what the elemental abundances were in the solar nebula. The answer works out at about 73 per cent hydrogen, 25 per cent helium, and about 2 per cent for all the rest. Of this, there is about 1 per cent oxygen, not quite as much carbon, approximately a tenth of a per cent of nitrogen, neon, iron, silicon, magnesium and sulfur, and about a hundredth of a per cent of argon, aluminium, sodium, calcium and nickel. The other elements are rarer still.

If you were to take a tonne of this material and cool it to the temperatures prevailing in the outer reaches of the solar nebula, you would end up with about 984 kg of hydrogen and helium gas, 11 kg of ice (not just frozen water, but other frozen 'volatile' substances such as methane, ammonia, carbon monoxide, and nitrogen), 4 kg of rock (mostly silicon plus oxygen, making, in combination with various metallic elements, **minerals** described as **silicates**) and a bit less than 1 kg of leftover metal (mostly nickel and iron). Closer to the Sun than Jupiter, it was too warm for ices to condense so the only solid particles in the nebula were rock and metal. Surprisingly, minerals of the kind that form by crystallization of molten rocks on Earth can (and did) form directly by condensation of their constituent elements from the gas of the solar nebula. Grains and globules that grew in this way can be identified within some kinds of meteorites. Working outwards from the Sun, water-ice would begin to condense at about the distance of Jupiter's orbit, but ices of other compositions would not appear until Saturn's orbit or beyond.

Some of the water now in the Earth's oceans was delivered late in the planet-forming process by comets that struck the Earth. However, most of it probably began as molecules of water trapped from the solar nebula within the structure of some kinds of silicate minerals, a process that was not affected by the 5 AU limit for ice condensation. This trapped water later escaped to the surface, along with other gases now found in the atmosphere, through volcanic eruptions.

Subsequent chapters show how these ingredients were put to use in building the planets. In many cases the composition of Solar System bodies, or at least of their atmospheres and surfaces, can be confirmed by **spectroscopy**. This technique relies on the fact that different substances absorb sunlight at specific wavelengths, so under favourable circumstances both the presence and relative abundance of a substance can be established by means of spectrum obtained through a telescope or by a spaceprobe. When a spectrum has not been obtained, even such a basic measurement as the proportion of incident sunlight that a surface reflects, a property referred to as **albedo**, can give useful information. For example clean ice reflects over 90 per cent of the light falling on it and so has an albedo of greater than 0.9, whereas carbonaceous material reflects less than 10 per cent and so has an albedo of less than 0.1.

Planetary heat

The heat generated during high velocity collisions means that planetary bodies formed by giant impacts and other collisions must have been very hot originally. Heat has been leaking out to their surfaces ever since, where it is radiated away to space. Larger planetary bodies lose heat proportionally more slowly than smaller ones, because the area of surface (over which heat is lost) increases with the square of a body's radius, whereas the volume (and therefore the amount of heat contained) depends on the cube of the radius. Even so, if the primordial heat inherited from the time of formation were the only source of heat, the interiors of most planetary bodies would have cooled down considerably by now. However, there are two additional sources of heat of great importance in keeping interiors warm, with dramatic consequences at the surface in the form of geological activity.

One of these heat sources, described as **radiogenic heating**, is the decay of radioactive isotopes (principally of uranium, thorium and potassium), which tend to be concentrated in the rocky parts. Radiogenic heating is responsible for at least half the heat leaking out of the Earth's interior today. Because the amount of radiogenic heating depends roughly on the volume of rock within a planetary body, it is the larger and rockier bodies that are most strongly heated by this mechanism. Within any body, the rate of radiogenic heat production decreases over time, as the proportion of remaining radioactive isotopes decreases. In the case of the Earth it has probably decreased fourfold in the past 4.5 billion years. Generally speaking, we would expect a planetary body's lithosphere to thicken over time, as the body as a whole ages and cools.

The other heat source, of tremendous importance for some of the satellites of the giant planets, is **tidal heating**. This is caused by the slight but continual tidal distortion of the shapes of bodies in close orbit around a much more massive body, and is independent of the size of the satellite itself.

Planetary facts

In the next chapter we will start our planet-by-planet tour of the Solar System with Mercury, the innermost planet, and work our way outwards in subsequent chapters. Each planet's chapter begins with a *planetary facts* list, where you can find that body's most important properties gathered together for easy reference. These properties and the units chosen to express them are explained below.

A body's size is expressed by means of its *equatorial radius*. This is the distance from its centre to a point on the equator (at the solid surface, except for giant planets where the surface is defined as the cloud tops). Large planetary bodies tend to be very slightly flattened towards their poles as a result of their rotation, so a body's polar radius (from its centre to one of the poles) is usually a little bit less. The difference is only about 20 km in the case of the Earth, but amounts to several per cent in the case of each of the giant planets.

Mass is expressed relative to that of the Earth. This is more convenient than expressing it in more familiar units. The Earth's mass is very nearly 6×10^{24} kg, which is 6 million million million million kg, or 6 thousand million million million metric tons.

Density is expressed in grammes per cubic centimetre (g/cm³), a unit that is particularly convenient because water (and ice) have a density of about 1 on this scale.

Surface gravity is expressed relative to that at the Earth's surface, where the gravitational force on an unsupported body is such as to cause it to accelerate downwards at a rate of 9.8 m s⁻².

Rotation period is defined by how long it takes the body to rotate relative to the distant stars. It is quoted in hours or days, whichever is more convenient. The 'day' used here is an Earth-day (24 hours), not the planet's own day length.

Axial inclination is used as defined on page 11 and is quoted in degrees. An inclination of 0° would mean that a planet's rotation axis was exactly perpendicular to its orbit.

Distance from the Sun is the body's average distance from the Sun, which is the same as half the length of the long axis of the ellipse traced by its orbit. Its value is expressed relative to the Earth's average distance from the Sun, i.e. in Astronomical Units (AU).

Orbital period is how long it takes the body to complete one orbit of the Sun. This is expressed either in days or years (the days and years being of course Earth-days and Earth-years).

Orbital eccentricity is quoted to show how eccentric (i.e. elongated) the orbit is, as defined on page 8. The lower the value, the more circular the orbit.

Composition of surface is listed to show at a glance whether the surface is rocky, icy or gassy, and the *mean surface temperature* is quoted to show how hot (or cold!) this surface is. In many cases, this mean value masks major differences in temperature between day and night and/or between poles and equator. Temperature is quoted in degrees centigrade (°C). If you prefer the Fahrenheit scale (°F): 0 °C is 32 °F, and 100 °C is 212 °F.

Composition of the atmosphere lists the four or five most abundant components in the body's atmosphere, and *atmospheric pressure at surface* is quoted relative to the Earth to give an indication of how dense it is.

Finally *number of satellites* indicates the number of satellites known at the time of going to press. Each of the four giant planets could have several other undiscovered satellites more than a few km in size.

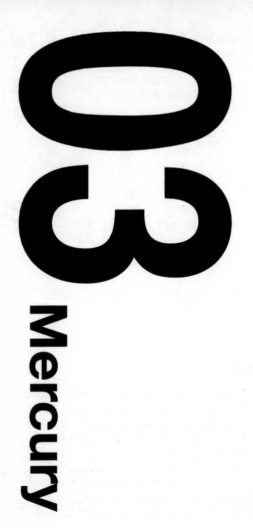

03

Mercury

In this chapter you will learn:
- about the curious relationship between the length of Mercury's day and its year
- how little we currently know about Mercury's intriguing surface and interior, and about plans for future space missions to remedy this.

Planetary facts	
Equatorial radius (km)	2439
Mass (relative to Earth)	0.055
Density (g/cm³)	5.43
Surface gravity (relative to Earth)	0.38
Rotation period	58.6 days
Axial inclination	0.1°
Distance from Sun (AU)	0.387
Orbital period	88 days
Orbital eccentricity	0.206
Composition of surface	rocky
Mean surface temperature	170 °C
Composition of atmosphere	atomic oxygen, sodium, helium, potassium
Atmospheric pressure at surface (relative to Earth)	10^{-15}
Number of satellites	0

Rotation and orbit

Mercury is the innermost planet, and its faster and shorter orbital path means that it overtakes the Earth every 116 days. It is best seen 22 days before or after these occasions, which is when it appears furthest from the Sun in our sky. This situation is described as maximum **elongation**. However, it is difficult to spot even then, because the two bodies are never more than 28° apart. Mercury's orbital eccentricity is 0.206, greater than any other planet except Pluto, and when maximum elongation coincides with Mercury's perihelion its angular separation from the Sun is only 18°.

Mercury's rotation period is exactly two-thirds of its orbital period, which means it rotates three times during the course of two complete orbits. If you puzzle out the effect of this 3:2 ratio as seen from Mercury's surface, you may be able to work out that the time from sunrise to sunset is exactly one Mercury-year (88 Earth-days). This means that Mercury's day is twice as long as its year!

The reason for this exact relationship between orbital period and rotation period must be related to the huge tidal influence of the Sun on such a nearby planet. Soon after it had formed Mercury was probably spinning much more rapidly than today, but tidal forces would have slowed down the planet's rotation to its present value within about half a billion years. Why the slow down did not continue until the rotation period was exactly one orbital period (88 Earth-days), in which case Mercury would always keep the same face towards the Sun, is a mystery. Similar tidal forces cause the Moon to keep the same face towards the Earth, and tidal-locking has had the same effect on the rotation periods of most satellites of the giant planets.

Missions to Mercury

Only NASA's Mariner 10 spacecraft has yet visited Mercury. This made three fly-bys of Mercury in 1974–1975, which allowed about 45 per cent of the planet to be imaged (Figure 3.1). Two further missions are now planned to this much-neglected planet. The first is a NASA probe named Messenger, launched in 2004. This is intended to fly past Mercury twice in 2008 and once in 2009, imaging the areas unseen by Mariner 10. Then in 2011 it should become the first probe to enter orbit about the planet, where it is expected to operate for about a year. A more ambitious follow-up is planned by the European Space Agency (ESA) which plans to dispatch a mission called BepiColombo in 2012. After a four year flight, the mission will deploy a sophisticated orbiter to obtain detailed images and a smaller independent orbiter operated by the Japanese Space Agency to study the magnetic field.

Temperature extremes and polar ice

Because Mercury is so close to the Sun, has such long days and nights, and has a more eccentric orbit than any other planet except Pluto, its surface is subject to great extremes of temperature. When Mercury is at perihelion, the surface temperature at the point where the Sun is directly overhead reaches a maximum of about 470 °C. Even when Mercury is at aphelion the noontime temperature is about 250 °C. In contrast, towards the end of Mercury's long night the temperature drops to as low as –190 °C. Furthermore, Mercury's axis of rotation is hardly tilted relative to its orbit so that the floors of craters near the planet's poles never see the Sun at all, and the local temperature is perennially very cold.

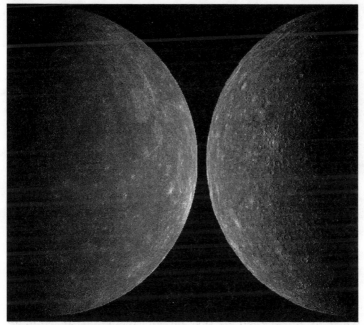

figure 3.1 Mercury as seen by Mariner 10 in 1974 on its incoming (right) and outgoing (left) fly-bys
there is a slight overlap in the areas covered in these two views
each view is constructed from a mosaic of more detailed images

One of the biggest surprises to emerge about Mercury since the time of Mariner 10 comes from Earth-based radar observations. These reveal very strong polarized radar reflections from the shadowed floors of Mercury's polar craters. The only reasonable explanation for this is the presence of ice. This ice is either at the surface or (more likely) dispersed within the shallow surface layer of fragmented debris known as **regolith** that covers all solid bodies lacking an atmosphere thick enough to act as a shield against meteorite impact. How the water to form this ice has been preserved on Mercury is a mystery. It seems highly unlikely that any water can have survived near Mercury's surface during the era of giant impacts when the planet was taking shape. It is more reasonable to assume that much of the polar ice is inherited from comets that have collided with Mercury during the 4.5 billion years since its formation.

The interior

Precise tracking of Mariner 10's trajectory as it passed the planet enabled Mercury's mass to be determined more accurately than previously possible, in the absence of any natural satellite of Mercury. This showed that Mercury is nearly as dense as the Earth. This is remarkable in so small a planet, because the Earth's high density is partly attributable to the compression of its interior resulting from its large mass. Mercury's high density, despite its small size suggests that it has an iron-rich core making up roughly 70 per cent of the planet's mass and 40 per cent of its volume. This means that the outer edge of the core is at a comparatively shallow depth, only about a quarter of the way between the surface and the centre. Relative to its total size, Mercury's core is much larger than that of any other terrestrial planet.

Current models for the condensation of the solar nebula cannot explain such a great preponderance of iron within Mercury as a direct consequence of the composition of the planetesimals and planetary embryos that collided to form the planet. Perhaps Mercury once had a thicker rocky mantle but this was blasted away by a late giant impact, rather like the one that hit the Earth to form the Moon, except that in Mercury's case the debris dispersed into space rather than clumping together to form a satellite around the planet.

We do not know whether Mercury's core is solid, or, like the Earth's core, a mixture of solid and liquid. On the one hand, a body as small as Mercury loses heat more efficiently than a larger body like the Earth, so we would expect its internal temperature by now to have fallen sufficiently far that the entire core would have solidified. On the other hand, Mercury has a magnetic field of sufficient strength (about a thousandth of the Earth's) to suggest that there remains a thin shell of molten material, perhaps an iron-sulfur mixture, surrounding the solid core. Currents within this liquid outer core could account for the origin of the magnetic field. However, without the stimulus of rapid planetary rotation, motion in the fluid part of the core ought to be too sluggish to produce a magnetic field as strong as the one observed. The magnetic field might therefore indicate what is known as 'remanent' magnetism in the core, which is magnetism 'frozen in' to the core when it solidified, just as toy magnets retain a magnetic field that has previously been imposed upon them.

The atmosphere

Mercury's gravity is too slight for it to be able to hold on to a gaseous envelope, so its atmosphere is tenuous in the extreme. Atoms that have been detected in Mercury's atmosphere are shown in the planetary facts list at the start of the chapter. These are all light enough to escape to space, and what has been detected must be a steady-state mixture that is continually replenished. Presumably these elements are either supplied continually by micrometeorites or are liberated from the surface under the influence of solar energy or meteorite impact.

Although water has not been detected in the atmosphere, it is likely that some or most of Mercury's polar ice derives from molecules of atmospheric water supplied by cometary impact or outgassing. These molecules would have wandered into the cold polar regions and become incorporated into the surface before they were able to escape or be split into hydrogen and oxygen atoms by radiation.

The surface

The Mariner 10 images of Mercury show a surface that appears deceptively Moon-like, though in fact the most densely cratered regions of Mercury bear fewer impact scars than the equivalent areas on the Moon, suggesting that overall Mercury has a slightly younger, but still very ancient, surface. The largest impact feature seen on Mercury is the Caloris basin (Figure 3.2) and is 1340 km across. This is described as a **multiringed impact basin**, because it consists of a series of concentric fractures, and is comparable in size with several similar features on the Moon. The object responsible for creating the Caloris basin was probably an asteroid about 150 km in diameter. It must have struck Mercury about 3.85 billion years ago, to judge from the number of smaller and younger craters that have been produced on top of it.

Much of the terrain in the upper right of Figure 3.2 shows the scars of ejecta that was flung out for a thousand km or more by the impact, but the effects of the Caloris impact were truly global in extent. The landscape on the part of the globe lying exactly opposite to the impact was severely disrupted by seismic shock waves that travelled right through the planet's interior and surface waves that travelled around the globe and converged on this most distant point (Figure 3.3).

figure 3.2 Mariner 10 view of the Caloris basin, covering an area about 1100 km across

only this half of the Caloris basin has been seen, because the other half was in darkness during each Mariner 10 fly-by

the Caloris basin occupies the centre left of the left-hand view in Figure 3.1

The large expanses of the floor of the Caloris basin that have not been overprinted by more recent cratering show extensive fracturing and ridging of the surface (Figure 3.4). This material is best interpreted as flows of **lava**, in other words, originally molten rock that welled up and filled the basin at some time after its formation, and indeed lava-filled depressions on the Earth have mixtures of ridges and fractures such as those seen (albeit at a much larger scale) on Mercury. Many of the ridges are compressional features where flows converged, or indicate inflation by injection of lava below a solidified carapace, whereas the straight-edged fractures indicate settling of the surface as the lava cooled and contracted. There are other areas of Mercury's surface that are relatively smooth and have few craters that are also probably covered by lava flows. On the Moon such areas are more obvious because the lava flows are dark, whereas on Mercury (for reasons unknown) the flows

figure 3.3 400-km wide Mariner 10 image showing part of the terrain lying on the far side of Mercury from the Caloris basin (i.e. near the right centre of the right-hand view in Figure 3.1)

the terrain was contorted by the converging seismic waves and flung up into a chaotic jumble of hills 5–10 km wide and up to 2 km high

the rims of pre-existing craters have been disrupted in this manner, but the younger craters, which post-date the Caloris impact, are entirely ordinary in appearance

have much the same brightness as the rest of the surface. However, Mariner 10 obtained some colour information (by combining images recorded through ultraviolet and orange filters) that reveals subtle colour contrasts between the supposed volcanic rock of the smooth plains and the impact debris that covers the more rugged ancient regions. Some of the volcanic regions have sharp edges, which is consistent with lava flows, but others have more diffuse boundaries and could possibly be deposits distributed by explosive volcanic eruptions. This is all very intriguing, and we desperately need another mission to Mercury capable of recording higher resolution (i.e. more detailed) images in multiple wavelengths to enable the origin

figure 3.4 detail overlapping the lower left of Figure 3.2, showing a 350-km wide area within the Caloris basin
ridges and fractures can be seen on the basin floor, as well as a number of younger impact craters

and composition of Mercury's surface to be better understood. One also wonders what might be lurking in the 55 per cent of the planet's surface that has not yet been imaged.

Global contraction

Apart from impact cratering, which must be continuing to the present day, the youngest event identified in Mercury's global record is the formation of a number of sinuous features known as lobate scarps (Figure 3.5). These range from 20 to 500 km in length, and up to 2 km in height. They are regarded as unmistakable signs of compression of Mercury's lithosphere, indicating where the edge of a tract of lithosphere has been thrust over an adjacent tract. Summing the deformation indicated by all the observed lobate scarps indicates a reduction in Mercury's radius of between 1 and 2 km. This could have been caused

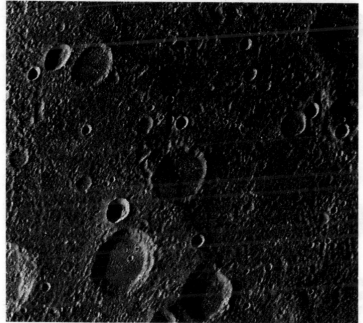

figure 3.5 200 km wide Mariner 10 image showing detail of an area near the upper centre of the right-hand view in Figure 3.1

a compressional scarp known as Santa Maria Rupes runs slightly obliquely from top to bottom of this view

it cuts through, and therefore post-dates, the 25 km diameter crater near the centre

either by contraction of Mercury's mantle as it cooled, or by solidification of a previously liquid part of the core.

This deformation of the surface appears to have been the 'last gasp' for Mercury's geology. Judging from the number of younger impact craters that are superimposed on the scarps, it must have happened several billion years ago. Mercury is too small a planet to have retained enough primordial heat, or to generate enough heat by radioactive decay, to have been geologically active since then. Furthermore, tidal heating appears to be slight. However, many other planetary bodies, among them Venus as described in Chapter 04, show abundant signs of geological activity continuing into more recent times and have substantial atmospheres with complexities of climate and weather that rival those of our own planet.

04

venus

In this chapter you will learn:
- what we know about Earth's sister planet
- about its dense atmosphere, its scorching hot surface and its weird geological history.

Planetary facts	
Equatorial radius (km)	6052
Mass (relative to Earth)	0.815
Density (g/cm³)	5.20
Surface gravity (relative to Earth)	0.90
Rotation period	243 days
Axial inclination	177.4°
Distance from Sun (AU)	0.723
Orbital period	224.7 days
Orbital eccentricity	0.007
Composition of surface	rocky
Mean surface temperature	480 °C
Composition of atmosphere	carbon dioxide (96%), nitrogen (3.5%), sulfur dioxide (0.015%), water vapour (0.01%), argon (0.007%)
Atmospheric pressure at surface (relative to Earth)	92
Number of satellites	0

Rotation and orbit

Venus is the planet that comes closest to us, passing within as little as 39 million km each time its faster, inner, orbit causes it to overtake the Earth, which happens every 584 days. Venus cannot be seen in the sky then, because it lies too close to the Sun and in any case the side facing us is not illuminated. However, Venus is clearly visible before and after this time (when a telescope will show it as a crescent shape), and is the brightest natural object in the sky apart from the Sun and Moon.

Venus has the most nearly circular orbit of all the planets, with an eccentricity of only 0.007. Because its orbit lies inside our own, Venus's position in the sky never strays more than 47° from the Sun. These times of maximum elongation occur 72 days before and 72 days after Venus overtakes the Earth. First Venus is visible in the evening sky after sunset, then as it begins to overtake us it draws closer to the Sun and becomes lost in the

figure 4.1 Venus as seen through a violet filter by the Galileo spacecraft in 1991 as it passed by Venus on its way to Jupiter
a blanket of sulfuric acid clouds obscures the surface, 60 km below
movement of the cloud tops is from right to left (east to west)

twilight, but then reappears as an equally bright object in the morning sky before dawn. Eventually Venus's faster orbital motion leaves the Earth so far behind that, as seen from Earth, Venus goes behind the Sun, to reappear later in the evening sky.

Despite Venus's proximity to Earth, even so basic a property as its rotation period remained unknown until the mid-1960s because it has a dense cloud cover (Figure 4.1) that hides the surface from view except by means of radar. Venus's 243 Earth-day rotation period was a surprise, the extreme slowness of this being one of the many aspects in which Venus differs from the Earth, which in size, mass and density is virtually Venus's twin. This rotation period (measured relative to the rest of the universe, not relative to the Sun) is longer than the planet's orbital period, so it is technically retrograde. If the Sun could be seen from the surface of Venus (though it probably never can, because of the cloud) it would rise in the west and set in the east. The day length on Venus is 116.7 Earth-days, so that the Venus-year consists of not quite two Venus-days.

Missions to Venus

After the Moon, Venus was the first planetary body to be targetted by space probes. The first, Venera 1, was launched by the Soviet Union in February 1961 and flew past Venus at a range of 100,000 km without returning any data. The Americans were more lucky in the following year with their Mariner 2 probe that flew past at 34,800 km and measured a pattern of microwave radio emission that showed that the

Name	Description	Date of fly-by or landing
Mariner 2	fly-by at 34,800 km; confirmed high surface temperature	Dec. 1962
Venera 4	parachute descent to 24 km; measured atmospheric properties	Oct. 1967
Venera 8	soft lander; measured atmosphere and surface properties	July 1972
Mariner 10	fly-by at 5875 km; images of cloud layer, atmospheric investigations	Feb. 1974
Venera 9	soft lander and orbiter; first pictures from the surface	Oct. 1975
Venera 10	same as Venera 9	Oct. 1975
Pioneer-Venus	radar mapping of surface from orbit (50 km resolution), ultraviolet imaging of clouds, atmospheric probes	Dec. 1978
Venera 11–14	soft landers; surface imaging and analysis	Dec. 1978–Mar. 1982
Venera 15 & 16	orbiters; radar mapping of northern hemisphere (1–2 km resolution)	Oct. 1983
Vega 1 & 2	soft landers and balloons; studied atmospheric composition and dynamics, also surface composition	June 1985
Magellan	orbiter; used radar to obtain detailed images and topographic data for the entire globe	Aug. 1990

table 4.1 some of the successful missions to Venus
Venera and Vega missions were launched by the Soviet Union, the others were by NASA

surface of the planet must be extremely hot. This was the beginning of the end for theories that beneath its cloud cover Venus might be a 'tropical' world covered by a global ocean or swampy jungle.

Venera 3, the first probe to impact another planet, entered Venus's atmosphere in March 1966 but failed to return any data. However, beginning with Venera 4 in October 1967 probes became both more ambitious and more successful. Some of the more noteworthy landers and orbiters are listed in Table 4.1, culminating with Magellan, a NASA mission that operated in orbit around Venus from August 1990 until March 1994. Magellan mapped 97 per cent of the surface by radar imaging (as illustrated by numerous figures later in this chapter) mostly with a resolution of about 100 m. Imaging radar operates by transmitting a radar beam onto the terrain to one side of the spacecraft's track and constructing an image by complex processing of the echos that bounce back to the antenna. For simple interpretation such images can be treated like photographs, except that slopes facing towards the radar beam appear brighter than slopes facing away from it, and that brightness also correlates with the roughness of the surface (rough surfaces appear brighter than smooth surfaces).

Magellan also used a radar beam directed straight downwards to measure the distance between the spacecraft and the ground, acting as a radar altimeter that mapped the global topography with a spatial resolution of between 2 and 20 km and a height precision of about 50 m. Once this phase was complete, the spacecraft was put into a lower circular orbit to enable spatial variations in the planet's gravity field to be examined by means of precise tracking of the probe's track. The mission ended in October 1994 when the orbit was deliberately lowered still further so that atmospheric drag could give information on the density of the upper atmosphere before the probe was destroyed by frictional heating.

Only one future mission to Venus has secured provisional funding yet. This is the European Space Agency's Venus Express (launch 2005), intended to probe the atmosphere and to study the surface using radar. The next phase of Venus exploration may be based upon 'aerobots', which are robotically-controlled balloons. Most of their time would be spent studying the atmosphere, but those designed to float below the clouds could also record optical images of the ground below. Some could be capable of making brief forays down to ground level to collect samples, and then escape back to the cooler high atmosphere before their electronic components became overheated.

The atmosphere

If a planet has an atmosphere this can exert very important controls on conditions at the surface. Venus has a much denser atmosphere than any other terrestrial planet, so thick that meteorites smaller than about 30 m do not survive passage through it, but burn up completely because of friction. In consequence there are no impact craters on the surface less than about 3 km across. This measure of protection of the surface against impacts must be offset against the twin forces of wind movement (which can redistribute surface materials) and 'weathering' whereby atmospheric constituents can react with and corrode the surface.

An atmosphere has an equally important role in determining the temperature at the surface. It precludes large day–night and equator-to-pole temperature extremes of the kind noted for Mercury, but it also raises the average temperature. If Venus had no atmosphere, its surface temperature would be nearly 500 degrees centigrade lower than it actually is (similarly, if the Earth had no atmosphere, its temperature would be about 30 degrees lower). This is because of the well-known, though not necessarily popularly understood, phenomenon of the **greenhouse effect**.

Essentially, in the greenhouse effect the atmosphere acts as a blanket keeping the surface warm. The global temperature is raised because solar energy, which peaks in the visible part of the spectrum, heats the planet's surface. The heated surface radiates heat away in the infrared. If there were no atmosphere the surface temperature would equilibrate at a level where the rate of energy absorption (in the visible) balanced the rate of energy loss (in the infrared). However, gases in the atmosphere absorb this infrared radiation very effectively, so the lower atmosphere is heated, which in turn keeps the surface warmer than it would otherwise be.

In the case of Venus, despite the cloud cover there is plenty of solar energy reaching the lower atmosphere. The high density of Venus's atmosphere alone would produce a stronger greenhouse effect than the Earth's atmosphere, but this is enhanced still further because carbon dioxide, the most abundant molecule in Venus's atmosphere, is a highly effective greenhouse gas, whereas nitrogen, the most abundant molecule in the Earth's atmosphere, is not. Temperature profiles through Venus's atmosphere by day and by night are compared with those for Earth and Mars in Figure 4.2.

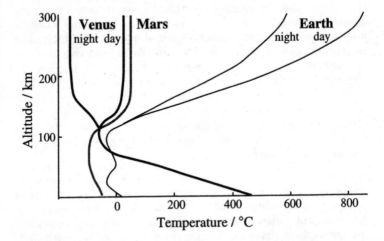

figure 4.2 the variation of temperature with altitude in the atmospheres of Venus, Earth and Mars
the very high temperature in Venus's lower atmosphere is because of the strong greenhouse effect
the high temperature in Earth's outer atmosphere (the 'thermosphere') is because of ionization of oxygen by solar ultraviolet radiation, and there is a smaller effect in Venus's thermosphere because of ionization of carbon dioxide
this occurs by day only, hence the day–night temperature differences
there is little or no such effect in Mars's atmosphere

Venus's atmospheric composition, shown in the planetary facts list at the start of the chapter, is dramatically different to the Earth's in its superabundance of carbon dioxide. However, this does not mean that Venus as a whole is richer in carbon dioxide than the Earth; the carbon dioxide that might otherwise be in the Earth's atmosphere has been removed by living organisms, and is locked up in deposits of limestone.

It is feasible that the young Venus and Earth had similar atmospheres, produced largely by volcanic degassing. However, their subsequent evolution was very different. Theory has it that Venus's closeness to the Sun led to an increased rate of evaporation of water into its atmosphere. This in turn increased the temperature because water vapour is a greenhouse gas, thereby leading to an even greater rate of evaporation and hence further temperature rise. This 'runaway greenhouse effect' evaporated any seas that Venus may once have had, and, without seas and living organisms to build limestone, Venus's carbon dioxide remained in the atmosphere.

If this really occurred, then why isn't Venus's atmosphere rich in water vapour too? The generally accepted explanation is that the strong solar ultraviolet radiation striking Venus's outer atmosphere splits molecules of water vapour into hydrogen and oxygen. This process is called **photodissociation**. The hydrogen is sufficiently light to escape to space, and there is compelling evidence that Venus has in fact lost a large quantity of hydrogen. This evidence is the fact that among what little hydrogen remains on Venus, deuterium, an isotope of hydrogen that is heavier than normal hydrogen (and which would therefore escape less readily) is much more abundant relative to ordinary hydrogen than in Earth's atmosphere. The fate of the oxygen liberated by photodissociation of water vapour in Venus's atmosphere seems to have been to react with rock at the surface to form oxides, because it too is extremely rare in the present atmosphere.

The clouds that render Venus's surface invisible are droplets of sulfuric acid (essentially a combination of water and sulfur dioxide), and extend from about 45 to about 65 km above the surface. Imaging of the clouds (Figure 4.1) enables the pattern of atmospheric circulation to be made out. It is apparent that most of the airflow at cloud top height is in the same direction as the rotation of the planet, but considerably faster so that the clouds circle the globe in only four Earth-days (corresponding to high-altitude winds of about 100 m per second). Within and below the cloud layer, warm air rises near the equator, flows poleward at high altitude, sinks groundward near the poles and returns towards the equator at low altitude, completing a circuit described as a **Hadley cell**. This pattern on Venus appears to be much simpler than on Earth, presumably because the Earth rotates much faster and has oceans, both of which profoundly influence atmospheric circulation.

Surface windspeeds on Venus are sluggish, typically only about one metre per second. Because the atmosphere is so dense, such winds are capable of transporting fine sand particles. However, these grains would have little abrasive effect upon collision with the surface, partly because of their slow speed (about a fifth of what it would be on Earth for wind to transport grains of equivalent size) and partly because the high surface temperature must reduce the propensity of rock to fracture under 'sand blasting' conditions.

The interior

Venus's density is only a little less than that of the Earth. This implies that its core probably occupies about 12 per cent of its volume, with the core-mantle boundary being about halfway between the centre and the surface. Venus has no magnetic field, but even if part of this core were liquid (as is the case in the Earth) we would not necessarily expect a magnetic field to be generated within it because Venus rotates too slowly to stir up the necessary currents. Unlike Mercury, there is no region of Venus's surface or interior cool enough to retain even a remanent magnetic field.

Debate continues about conditions within Venus's mantle, in particular whether or not it has degassed to the same extent as the Earth's. If it has not, then that would be an additional factor contributing to the scarcity of water in Venus's atmosphere. The thickness of the lithosphere is also open to question. Mapping of Venus's gravity field by Magellan gives some clues as to the outer structure of Venus. In particular it seems that the average crustal thickness is 20–40 km, and that this crust is the top part of a lithosphere that is up to 100 km thick. Furthermore, some topographically high regions seem to be supported from below by upwelling of hot mantle below the lithosphere.

Unfortunately, the interpretation of gravity fields is always ambiguous, and alternative models can be made to fit the same data. What we really need is to establish a network of seismic stations on Venus's surface, that would record the vibrations triggered by 'earthquakes' and enable a picture of the planet's internal layered structure to be built up, by deducing the paths followed by these vibrations through its interior. So far this has been achieved only for Earth and the Moon.

Global topography

Whatever the shortcomings of our knowledge of Venus's interior, at least its surface is now known quite well. The global topographic mapping of Venus by Magellan is illustrated in Plate 1, although at the scale of this plate much of the detail is lost. The colour-coding used to indicate surface height makes the planet look misleadingly Earth-like, with blue (low) areas corresponding to oceans and green and brown areas (high) resembling continents. However, on the Earth there is a fundamental difference between the deep ocean floors and the continents, quite apart from the fact that the former are flooded by water (which is absent on Venus). This is that surface height distribution on the Earth is distinctly bimodal; in other words when a plot of the proportions of the Earth's surface at different heights is made there are two distinct concentrations, one corresponding to the ocean floors (at 4–5 km below sea-level) and the other representing the continental plains at just above sea-level. The reason for this is that the Earth's oceans and continents are floored by two distinct types of crust: high-density low-lying oceanic crust and low-density upstanding continental crust. In contrast, a similar plot for Venus shows just a single peak clustered around the average planetary radius of 6051.8 km, and in fact nearly two-thirds of Venus's surface area lies within 500 metres of this mean elevation. Here is compelling evidence that Venus lacks the clear distinction between Earth-like oceanic and continental crust.

However, the topography of Venus is Earth-like in other ways, notably in its mountain belts. The highest mountain region is Maxwell Montes, which reaches 12 km above the mean elevation and can be seen in Plate 1 above right of the centre of the western hemisphere. On high ground such as this, the lower temperature would favour reaction of atmospheric gases with surface rocks to generate carbonates, sulfates and sulfides that would be unstable if transported to lower altitudes. Therefore it is unsurprising that Magellan's radar instrument revealed that Venus's mountain tops have electrical properties consistent with a frosting of iron sulfide.

The surface

Soviet landers recorded images on the surface (e.g. Figure 4.3) at four sites, all just to the right of the centre of the western

figure 4.3 part of Venera 13's view of the surface of Venus
the surface is covered by broken slabs of rock, which is typical of the
landing sites for all four successful landers that had cameras
the saw-tooth edge is part of the lander, and the bright semicircular object is
the discarded camera cover

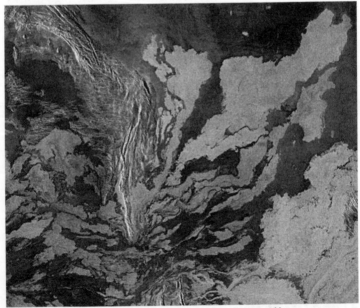

figure 4.4 400 km wide Magellan image of lava flows on Venus
(47° S, 25° E)
the flows that appear very bright to the radar must have rough surfaces,
whereas the darker flows are evidently smoother

hemisphere of Plate 1, and made crude determinations of surface composition at five sites (the four imaged sites plus the Vega-2 landing site just south of the equator at 179° W). These indicate composition similar to volcanic lava of the type known as **basalt**, though it must be borne in mind that analyses at so few sites are not necessarily representative of the surface in general. Soviet orbiters also recorded radar images of the surface, but these covered only the northern hemisphere and our view was revolutionized by the much more detailed and global imaging radar coverage by Magellan.

Venus's generally volcanic nature as suggested by the analyses and images made on the surface is confirmed by Magellan images such as Figure 4.4, which show unmistakable lava flows. In this instance the lava flows have been fed from a source some 300 km to the west, have breached the north-south ridge belt near the centre of the image and spread across the plains beyond. The freedom with which the lava flows have spread is consistent with a generally basaltic composition, basalt being the lowest viscosity (most runny) kind of lava that is common on the Earth today.

All low-lying regions on Venus appear to have been flooded by lava, though usually the boundaries between individual flows are not so clear as on Figure 4.4. Even the high mountains may originally have been volcanic plains that were deformed by folding and thrusting. The apparently oldest landscapes on Venus are tracts of intensely fractured land (described as tessera terrain) that peep out through the lava plains here and there, and are presumably remnants of formerly much more extensive regions that have been buried by lava. Tessera terrain now occupies only 8 per cent of Venus's surface. Examples of all three terrain types occur together in Figure 4.5.

Venus has many other obviously volcanic features in addition to the lava flows. These include a great many large gently sloping volcanoes of the kind described on Earth as **shield volcanoes** and typically produced by the eruption of basalt lava through a persistently active vent. One such shield volcano on Venus is shown in Figure 4.6. The terrain in the foreground is fractured volcanic plains, which is the oldest unit in this area. This can be deduced because the plains can be seen to be buried by the lava flows that have descended from the volcano, and also by the ejecta from the impact crater.

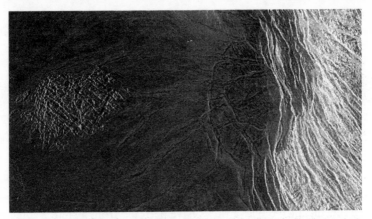

figure 4.5 240-km-wide Magellan image (65° N, 0° E) showing the edge of the radar-bright Maxwell Montes, the highest region on Venus (right) the volcanic plains to its west contain a surviving 60-km-long piece of tessera terrain that juts out, island-like, above them

figure 4.6 three-dimensional perspective view of the 5-km-high shield volcano Maat Mons (1° N, 194° E) and the adjacent region, made by combining imaging and altimetric data from Magellan the vertical scale has been exaggerated tenfold (to show the topography more clearly), so in reality slopes are much more gentle than implied here the impact crater in the right foreground has a diameter of 23 km

Although many of the features in Venus's landscape can be recognized by comparison with the Earth or other planetary bodies, one category that seems to be unique to Venus is **coronae** (singular: corona). These are approximately circular or elliptical features mostly a few hundred km in diameter marked by a pattern of concentric fractures, such as the example seen in Figure 4.7. However, they are clearly not multiringed impact basins because they are not circular enough and lack surrounding blankets of ejecta. The interior of a corona is usually domed upwards, though there may be a depressed 'moat' around this. There are often many small volcanoes and lava flows within a corona, and in addition to its concentric fractures a corona is often situated on a swarm of parallel

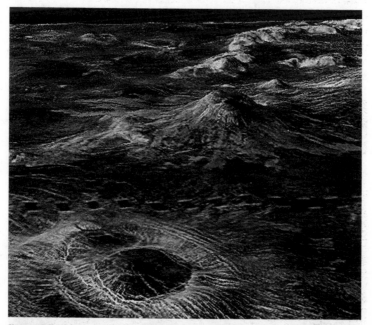

figure 4.7   three-dimensional perspective view (constructed in the same way as Figure 4.6 but covering a greater area) showing in the foreground a 200 km corona named Nagavonyi

note the central uplift surrounded by a ring-shaped depression and both concentric and linear fractures

behind and to the right is a 2-km-high shield volcano

the line of small distorted squares running across the view is caused by gaps in the radar image coverage

fractures that may extend across the adjacent terrain for several hundred km.

Coronae are believed to form above sites where hot material has risen through the mantle (a **mantle plume**). It is easy to see how this can explain the volcanism, but it can also explain the topography. A mantle plume would heat the crust, which would therefore expand and become uplifted and fractured. As the plume decayed, the heated crust would cool and therefore subside, further encouraged to do so by the fresh load of volcanic rocks upon it. This would add to the fracture pattern and cause the moat-like depression. Over 300 coronae have been identified on Venus, in all stages of evolution and preservation including features that could be incipient coronae that never quite developed, coronae that appear to still be dynamically supported above active mantle plumes, and decayed and fully subsided structures.

The volcanoes and coronae on Venus are not randomly distributed, but are concentrated near the north-south zone of high terrain in the western hemisphere (Beta Regio) and near the east-west belt of high terrain in the eastern hemisphere (Ovda

figure 4.8 130-km-wide Magellan image showing a 2-km-wide sinuous channel crossing fractured lava plains (49° S, 273° E)

Regio-Atla Regio). The gravity data indicate that these regions are supported from below because they overlie hot upwellings within the mantle. Fractures parallel to the crests of these rises attest to the stretching of the updomed crust. However, there are no concentrated alignments of volcanoes like those that indicate boundaries between tectonic plates on the Earth.

Another peculiarity of Venus's surface is shown in Figure 4.8. This is one of about 200 sinuous channels on the plains of Venus, the longest of which has been traced for an amazing 6800 km. Superficially these resemble meandering river channels, but generally speaking they lack tributaries. Given that Venus's surface temperature is likely to have been hot for a long time past, they are most unlikely to have been cut by flowing water. They could be channels cut by basalt lava, or some other, less viscous, variety of silicate magma. However, the erosive power of such lava, and its ability to remain molten for long enough to travel the length of a channel have both been questioned. An alternative possibility is that they were carved by flows of molten carbonatite, a lava composed largely of calcium

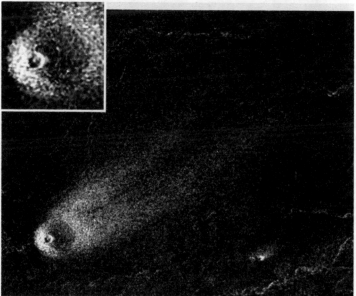

figure 4.9 90-km-wide Magellan image showing a small volcanic cone with a wind streak extending away from it towards the northeast (9° S, 247° E) note also the volcanic crater at the summit of the cone, best seen in the enlargement at the top left

carbonate. Because carbonatite does not solidify until it cools to below 600 °C, it could remain liquid for a long time after eruption onto the hot surface of Venus. Whatever the origin of these channels they all appear to be rather old. They are cut by many younger fractures and the landscapes across which they wind have been warped and tilted so that there is no longer a consistent downslope direction along each channel bed.

Although Venus lacks credible signs of flowing water, it shows plenty of evidence of another surface process familiar on Earth, namely wind. There are fields of dunes each up to about 5 km in length, and also 'wind streaks' on the surface that appear to correspond with the prevailing wind direction and extend downwind from topographic features. Figure 4.9 shows an example of a wind streak extending from an isolated volcano. This is most likely to be radar-bright (rough) material exposed by stripping away of a regional dust cover by turbulent vortices imposed in an otherwise gentle wind flow by the obstructing volcano.

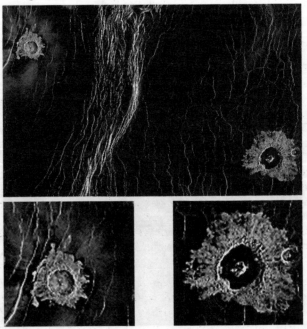

figure 4.10 top: 450-km-wide Magellan image showing fractured volcanic plains bearing two obvious impact craters surrounded by radar-bright ejecta (60° S, 206° E)
the craters are Eudocia (28 km diameter, top left) and Morisot (55 km diameter, lower right) below: enlarged view of each crater

About a thousand impact craters have been identified on Venus, the largest being 280 km in diameter. Because the dense atmosphere prevents small impactors reaching the surface with sufficient speed to form craters, the smallest impact craters on Venus are about 3 km across. Many of these are poorly formed or occur as overlapping clusters, suggesting the impactor broke up in the atmosphere shortly before impact. However, from 30 km diameter upwards each crater is virtually a work of art, with a central peak, sharp rim and a surrounding blanket of rough ejecta with a lobate edge characteristic of ejecta flow in a dense atmosphere. Two such craters appear in Figure 4.10. The radar-bright halo beyond the even brighter ejecta blanket surrounding the crater Eudocia in the upper left is more finely dispersed ejecta. Several of Venus's craters show flow-like features emerging from beneath the fringes of their ejecta blankets. An example of this is the radar-bright material to the left of the crater in the right foreground of Figure 4.6, which might be a pool of now-solidified melt produced by the energy of the impact.

Global history

With craters being so widespread on Venus there is plenty of opportunity to use the cratering timescale to determine the age of its surface. By counting the number of craters per unit area, the average age of Venus's surface works out at about 500 million years (which is half a billion years, about a tenth the age of the Solar System). It is only to be expected that Venus is a much more vigorous and active world than its smaller cousin Mercury. What is really surprising, however, is that Venus's craters are distributed entirely randomly across all terrain types, with no discernible difference in crater density between different types of terrain. This applies even in cases where one terrain unit, such as the volcanic plains, clearly overlies, and so must be younger than, another such as regions of tessera terrain.

This means that all terrain types must date from roughly the same era, about 500 million years ago. Nothing can be identified as having survived from before that time, nor are there any large expanses of surface that have been created since then. This does not mean that there has been absolutely no volcanic activity at all during the past half billion years, merely that the vast majority of the features we can see were created during an epoch of intense activity over a relatively brief period of time (lasting perhaps 50 million years). This period of time, culminating about 500 million years ago, is too short to resolve using crater counting.

The lack of clear signs of current or, geologically speaking, recent activity on Venus prompts the question of how Venus is able to lose its heat. Unless Venus is made of radically different material to the Earth, which seems inconceivable, then radiogenic heat must be being generated within Venus at much the same rate as within the Earth, which is only slightly larger. Most of the Earth's internal heat production escapes to the surface, and thence to space, by means of the creation of new, hot lithosphere and reabsorption of old cold lithosphere into the deep mantle through the action of **plate tectonics** (described in the next chapter). Heat escapes from the Earth's interior at a lesser rate through volcanoes. Venus shows no signs of plate tectonics and has few, if any, active volcanoes. Some sites of mantle upwelling are identifiable as the dynamically supported highlands, and regions of crustal compression presumably indicate underlying downwellings, but there is little or no on-going creation or destruction of lithosphere. The only way remaining for internal heat to escape is by conduction through the lithosphere, but this is too ineffective to allow the rate of heat escape to match the assumed rate of heat production.

This has led planetary scientists to speculate that Venus's interior is currently warming up. In particular, it seems that the part of the mantle immediately below the lithosphere must slowly be getting hotter because of the build-up of heat trapped below the lithospheric lid. As this sub-lithospheric mantle becomes hotter, it must expand and eventually become less dense than the overlying lithosphere. This could lead to catastrophic overturning, in which either the whole lithosphere or just the mantle part of it founders. In either case the globe would be resurfaced while any surviving previous crust would become deformed and flooded by magmas displaced from the hot interior. Subsequently, a slowly increasing thickness of rigid mantle would once again accrete beneath it.

Given that this catastrophic overturn is a predicted consequence of Venus's present condition, then maybe the same thing happened 500 million years ago, which would explain the relatively uniform age of Venus's current surface. This, in broad terms, is in fact the most widely accepted model for Venus's long-term behaviour. For periods of several hundred million years its surface sees little change, but, because conduction of heat through the lithosphere is unable to keep pace with the rate of radiogenic heat production, eventual global catastrophe is inevitable when the stored up heat escapes in an orgy of volcanism and lithospheric recycling before the situation stabilizes again.

To put the behaviour of Venus and other planetary bodies into perspective, the next chapter considers the Earth as a planet.

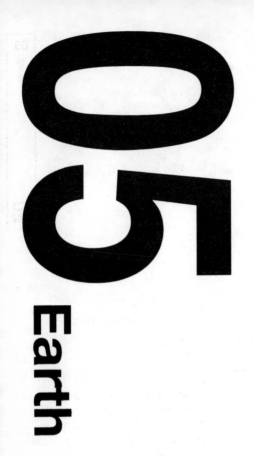

05

Earth

In this chapter you will learn:
- about the Earth in its context as one planet among many
- how both similarities and differences occur between the Earth and its most Earth-like companions.

Planetary facts

Equatorial radius (km)	6378
Mass	5.97×10^{24} kg
Density (g/cm³)	5.52
Surface gravity	9.78 m s^{-2}
Rotation period	23.93 hours
Axial inclination	23.45°
Distance from Sun	149.6 million km (= 1 AU)
Orbital period (days)	365.24
Orbital eccentricity	0.017
Composition of surface	rocky
Mean surface temperature	15 °C
Composition of atmosphere	nitrogen (78%), oxygen (21%), water vapour (1%), argon (0.93%), carbon dioxide (0.035%)
Atmospheric pressure at surface	10^5 kg m^{-2}
Number of satellites	1

The Earth is naturally the planet we know most about, because we live here and have a long record of detailed observations. Even so, some important aspects of the Earth's geology and atmospheric behaviour did not become apparent until it became possible to study it from the vantage point of space, using techniques similar to those upon which most of our knowledge about other planetary bodies is based. This chapter reviews the Earth as a planet, to put it into context with its fellows.

Rotation and orbit

The Earth rotates not, as you might think, in exactly 24 hours, but in 23 hours 56 minutes. This is because our definition of a day (which we divide into 24 hours) is based, naturally enough, by the average time between successive noons. This is the 'solar day' and is slightly longer than the Earth's true rotation period because the Earth's orbital motion causes the direction towards the Sun to change by about one degree over the course of a day. The true rotation period, relative to the stars, is 23 hours 56 minutes, and because of this any particular star will appear to rise four minutes earlier each day.

The Earth's 23° axial inclination is responsible for our seasons. When the northern end of the axis is tilted towards the Sun, the northern hemisphere experiences summer, and the southern hemisphere is in winter. Six months later, the situation is reversed and it is winter in the northern hemisphere and summer in the southern hemisphere. The axial inclination and the direction in space towards which the axis points both change very slowly. Currently, the axial inclination varies between 21.8° and 24.4° over a 40,000 year period, and it takes 22,000 years for its direction to describe a circle in the sky – a phenomenon described as **precession**. Precession of the Earth's axis means that in 11,000 years the seasons will have moved round the calendar by six months. These two cycles cause variations in the distribution of solar radiation falling on the Earth. The total amount of sunlight reaching the Earth is affected by changes in the eccentricity of our orbit, which varies from virtually zero to about 0.05 over a 110-thousand-year cycle. This combination of variations in axial inclination, precession and orbital eccentricity is believed to act as triggers for certain kinds of climate change such as ice ages.

The atmosphere

The Earth's present atmosphere is very different to that of Venus. Part of the reason is that the Earth's greater distance from the Sun prevented a 'runaway greenhouse effect' from occurring and boiling away all the oceans. A second important factor is the influence of life. Without plant life to break down the carbon dioxide (liberating oxygen and storing the carbon in organic compounds), there would be little or no free oxygen in the atmosphere. The oxygen content of the atmosphere had probably risen to something like its present level by about 500 million years ago and has not dipped much below that ever since. However, there is some geological evidence that it may have risen to as much as double the present amount for a few tens of millions of years about 300 million years ago. On the other hand, carbon dioxide was probably at about 100 times its present concentration in the atmosphere 2 billion years ago and had fallen to about ten times its present value by about 500 million years ago.

Oxygen molecules sufficiently high in the atmosphere to be exposed to solar ultraviolet radiation tend to become photodissociated into atoms that can then react with a surviving oxygen molecule to form ozone (a molecule containing three oxygen atoms, rather than the two atoms that make up the oxygen molecules we breathe). This 'ozone layer' is very insubstantial, and the total amount of ozone it contains would make a layer only a few millimetres thick if gathered together at sea-level. Even so, it is highly effective at blocking out ultraviolet radiation that would be harmful to life were it to reach the surface, and the partial destruction of the ozone layer by reaction with industrially produced chemicals is a persistent worry.

As noted in the previous chapter, Earth's atmosphere produces a modest greenhouse effect, amounting to about 30 degrees of warming, thanks chiefly to carbon dioxide and water vapour. The amount of greenhouse warming must have decreased over time as the carbon dioxide level fell. Astrophysical models suggest that the Sun's luminosity must have increased slowly by about a quarter since the birth of the planets. However, the gradual decline of the greenhouse effect seems to have compensated nicely for this, at least during most of the past 4 billion years for which sedimentary rocks have been preserved. These include deposits formed by water flowing across the land surface, proving that the temperature has been neither above boiling point nor (at the equator) freezing cold at any time during this period.

Life began on Earth at least 4 billion years ago, possibly beside hot underwater springs. It has been suggested, as part of the 'Gaia hypothesis', that life acts as a regulator on the Earth's temperature, keeping it within tolerable limits by adjusting the balance of gases in the atmosphere. However, there are growing signs of a human-induced global warming upsetting any natural balance by industrial and agricultural release of greenhouse gases such as carbon dioxide and methane.

The Earth's atmospheric circulation is more complex than that of Venus. Earth's much faster rotation controls the broad pattern, preventing the Hadley cells that rise at the equator extending beyond about 30° north and south. There is a polar Hadley cell in each hemisphere at high latitudes, which (like the tropical Hadley cells) is driven by the pole-to-equator temperature gradient. There is an intervening cell in each hemisphere, in which low altitude winds flow polewards (contrary to the temperature gradient at the surface) and high altitude winds flow equatorwards. The rising limb of this cell is driven by the rising limb of the polar Hadley cell and the descending limb is driven by the descending limb of the tropical Hadley cell. This basic Hadley cell pattern is sketched in Figure 5.1.

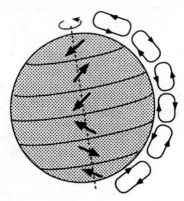

figure 5.1 Hadley cells in the Earth's atmosphere, showing the directions of airflow in cross-section (height greatly exaggerated) and the directions of prevailing surface winds in each latitudinal zone

the high-altitude winds are almost but not quite opposite in direction to the low-altitude winds, with the effect that the net airflow is from west to east round the globe

the three-dimensional shape traced by a parcel of air as it circulates in a Hadley cell is thus a loose spiral wrapped latitudinally around the globe

Your own experience will show you that prevailing surface winds do not actually blow north or south as this simple description would suggest. This is because the Earth's rotation deflects poleward flowing air to the east and equatorward flowing air to the west. This effect is responsible for such features as the 'northeast trade winds' in the Atlantic at the latitude of west Africa (the deflected low-altitude part of the northern hemisphere tropical Hadley cell), and the prevailing southwesterly airstreams that impinge on the Atlantic coast of Europe (the deflected low-altitude part of the adjacent Hadley cell). The poleward flowing air in the top of the northernmost Hadley cell is deflected particularly strongly to the east, and constitutes the famous 'jet stream' that makes west-to-east flights across the north Atlantic an hour or so faster than east-to-west flights.

Superimposed on the overall Hadley cell pattern are smaller scale rotating weather systems (e.g. Figure 5.2), which are another consequence of the Earth's rapid rotation. There are also patterns caused by the distribution of land and sea, and seasonal variations. We know that, in addition, the climate is not stable.

Some climate change is driven or triggered by combinations of the varying inclination and direction of the Earth's axis and the eccentricity of its orbit. Other climate change depends on the mean global temperature, which controls the energy available for storms, for example.

figure 5.2 Hurricane Floyd imaged by a weather satellite in Earth orbit on 14 September 1999, two days before it hit the Carolina coast of the USA, which was one of the most powerful tropical storms of the twentieth century, with windspeeds exceeding 200 km per hour
the area shown is 2700 km across, and part of the island of Cuba can be made out in the lower left

The interior

Our understanding of the Earth's interior is based largely on the paths followed by seismic waves, which are vibrations triggered by earthquakes or large explosions. Figure 5.3 shows how the core is detected by the refraction (bending) of wave paths as they pass from the mantle to the core. Seismic waves can also be reflected from boundaries between layers. By combining information from reflection and refraction we can detect boundaries such as the base of the crust (25–90 km in continental areas but only 6–11 km in oceanic crust), and the presence of an inner core (radius 1215 km) within the 3470 km radius outer core shown in Figure 5.3. Furthermore, we can tell that the outer core must be liquid, because although pressure waves pass through it, another kind of seismic wave involving shearing (or shaking) does not.

Earth's strong magnetic field is generated by motion within this liquid outer core, whose density suggests that it is composed of a molten mixture of iron and a less-dense element such as sulfur. The solid inner core is probably mostly iron with a few per cent of nickel.

The mantle has the composition of the rock type known to geologists as **peridotite**, which is presumed to constitute the mantle of the other terrestrial planets too. There are two quite distinct kinds of crust. The crust forming the ocean floors has

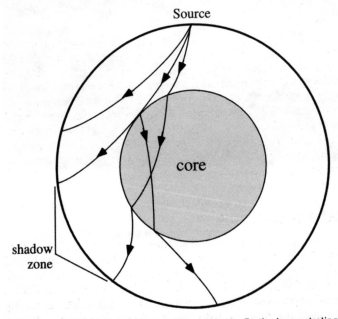

figure 5.3 paths of seismic waves deep within the Earth, demonstrating the presence and exact size of the core

the core's density is about twice that of the overlying rocky mantle, which halves the speed of the seismic waves

waves reaching the core-mantle boundary are deflected sharply downward, and there is a 'shadow zone' at a certain distance from the source within which seismic stations cannot pick up directly-transmitted waves

the composition of basalt (slightly richer than peridotite in silicon and oxygen), whereas continental crust is even more distinct from peridotite and consists of a range of rock types including andesite and granite. These are **igneous** rocks, meaning that they have crystallized from a molten state, either near at surface (in which case they are described as volcanic) or at depth (intrusive). The upper part of the continental crust has large amounts of **sedimentary** rock (accumulations of reworked fragments derived by erosion of pre-existing rocks), and lower down much of it has been deformed and recrystallized by heat and pressure to produce **metamorphic** rock. Oceanic crust is both thinner and denser than continental crust, so its surface is lower and virtually all oceanic crust lies well below sea-level (mostly at 4–5 km). The surface of most continental crust lies above sea-level, though it may be flooded to a depth of a few hundred metres at the edges of continents, especially where the crust has become thinned by stretching that precedes the opening of an adjacent ocean.

The oceanic crust is produced directly from melting in the mantle, and does not usually survive more than about 200 million years before being destroyed by the action of plate tectonics, as will be described shortly. Continental crust is largely a product of melts derived from the destruction of oceanic crust. The oldest traces of continental crust are about 4 billion years old, and it had probably grown to between 40 and 80 per cent of its present volume by 2500 million years ago. Volcanic activity causes the continental crust to continue to grow today, though only very slowly.

Plate tectonics

The Earth's lithosphere is broken into a number of plates, some surfaced only by oceanic crust, others carrying both oceanic and continental crust. The motion of these plates is described as plate tectonics, and provides the principal mechanism by which Earth's internally generated radiogenic heat is able to escape. This occurs by creation of hot new oceanic crust at boundaries where plates are moving apart, and destruction of old cold oceanic crust at boundaries where a plate is destroyed by descent below either a continental or oceanic edge of an adjacent plate. A crucial distinction with Venus is that plate tectonics on the Earth enables the rate of short-term heat loss to balance that of heat production. Thus, we do not conceive of catastrophic events for the Earth of the same magnitude as the half billion year lithospheric overturn proposed for Venus, although the geological record on Earth does show episodes of enhanced eruption of large expanses of basalt during eras when large continents are rifting apart.

Figure 5.4 shows the basics of plate tectonics in cross-section. The full thickness of the lithosphere is shaded, without distinguishing its compositional division into crust and mantle.

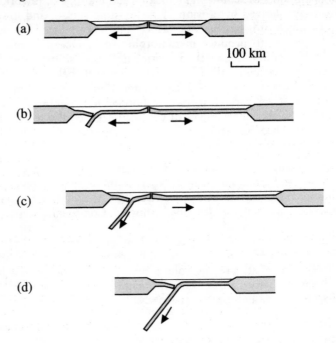

figure 5.4 time series of cross-sections to show the evolution of an ocean, over a span of a few hundred million years

in (a) the lithosphere is divided into two plates, separated by a constructive boundary where new material is added to the lithosphere as the plates diverge

in (b) the left-hand plate has broken into two, with its oceanic part beginning to descend (be 'subducted') below its continental part at a destructive plate boundary, so there are now three plates

in (c) the constructive plate boundary itself is about to be subducted

in (d) this has happened: only two plates remain and subduction will cease when the two continents collide, fusing the two plates together

currently the Atlantic ocean is in stage (a), though it is wider than shown here, and the Pacific Ocean (also much wider) is in stage (c)

The unshaded area below each cross-section is the asthenosphere part of the mantle, which is the weak zone across which the plates move. The sequence shows an ocean spreading apart between two formerly-united continents. Eventually a new plate boundary forms where one old, cold edge of the oceanic part of a plate is forced below the edge of the adjacent continent. Magmas generated by melting as this slab descends feed volcanoes (not shown) above. The ocean now begins to close, and will eventually be destroyed, bringing the two continents into collision.

The Earth's tectonic plates are in continual motion, and the rate of plate creation in one part of the globe matches the rate of plate destruction elsewhere so that the Earth's total surface area remains constant. The present-day global situation is shown in Figure 5.5. In addition to their topographic and geologic expression, plate boundaries can also be recognized because this is where most earthquakes occur. Note that in addition to constructive and destructive plate boundaries, there is a third type, conservative plate boundaries, where adjacent plates slide past each other. The speed of plate motion (which is similar to the rate of convective flow in the asthenosphere, though the two are not directly related) is typically a few centimetres per year.

Figure 5.5 also shows how most of the Earth's active volcanoes occur along destructive plate boundaries, these volcanoes being near the edge of a plate that is riding over the subducting part of another plate. Volcanoes not obeying this rule include those along the east African rift, where Africa is showing signs of splitting apart, and Hawaii, which sits above a mantle plume that rises from great depth and has left a trail of volcanoes as the Pacific plate has moved across it.

The Earth is a very dynamic planet, with a relatively young surface. Most traces of the Earth's earliest past have been obliterated by creation and destruction of crust through the action of plate tectonics together with erosion and deposition of sediment caused by wind, water and ice. These processes continually erase the record of craters formed by impacts, and surviving impact craters are therefore rare. However, our smaller neighbour, the Moon, has an arid, airless surface that has long been geologically almost inert and therefore abounds in impact craters. This is the subject of the next chapter.

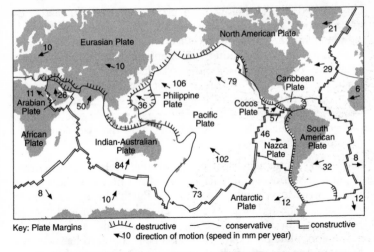

Key: Plate Margins ⊥⊥⊥⊥ destructive ⎯⎯ conservative ▭▭ constructive
◄—10 direction of motion (speed in mm per year)

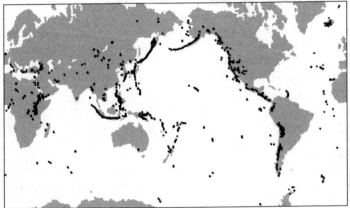

figure 5.5 the global distribution of plate boundaries (top) and volcanoes (bottom)

06

the Moon

In this chapter you will learn:
- about the nearest planetary body to the Earth
- about the long record of impact cratering on its surface, and about the ancient eruptions that flooded many low-lying areas.

Planetary facts	
Equatorial radius (km)	1738
Mass (relative to Earth)	0.0123 (7.35×10^{22} kg)
Density (g/cm³)	3.34
Surface gravity (relative to Earth)	0.17
Rotation period	27.32 days
Axial inclination	6.67°
Distance from Earth	384,400 km
Orbital period	27.32 days
Orbital eccentricity	0.05
Composition of surface	rocky
Mean surface temperature	1 °C
Composition of atmosphere	argon, helium, atomic oxygen, sodium, potassium
Atmospheric pressure at surface (relative to Earth)	2×10^{-14}
Number of satellites	0

Rotation and orbit

The Moon is a familiar object in our sky, because it is our nearest neighbour and the only natural object to orbit the Earth. As described in Chapter 02, this situation is thought to have arisen as the result of a giant impact, 4.5 billion years ago. Something that adds to the Moon's familiarity is that it always presents the same face towards us. This is because the Moon rotates exactly once per orbit, which is described as **synchronous rotation**. This is a phenomenon also exhibited by most of the satellites of the giant planets, and is brought about because tidal forces between the planet and its satellite slow the satellite's rotation until it matches its orbital period.

Although the Moon rotates once per orbit, with patience it is possible to see about 59 per cent of its surface from the Earth. One reason is that the Moon's proximity is such that the Earth's rotation results in slight but significantly different points of view from a single site across a 12-hour period. A similar change in perspective could be gained by moving from the north pole to the south pole. The other reason is that although the Moon rotates at an exactly constant rate, its progress round its orbit varies slightly in accordance with Kepler's second law of planetary motion (Chapter 02). However, the terrain near the edge of the lunar disc is never well displayed, because we only ever see it obliquely, and there remains 41 per cent of the lunar surface that can never be seen from the Earth.

Missions to the Moon

The Moon was the first extraterrestrial target for space missions. Probes have been directed towards it since almost the very dawn of the space age (Table 6.1), and it was the main focus of the 1960s–1970s 'space race' between the USA and the then Soviet Union. In the end, only NASA attempted to put people on the Moon, and the six successful Apollo landings brought back a total of 382 kg of lunar rocks. These samples, together with 0.3 kg collected from other sites by unmanned Soviet sample return missions, are what has enabled us to calibrate the cratering timescale and were immensely important in developing our current level of understanding of the Moon's origin and history.

The budget for the Apollo programme was terminated in 1972, after which there was little further effort in lunar exploration until 1994 when the Clementine probe went into lunar orbit and collected a wealth of previously unknown information about the topography, crustal thickness, and variations in crustal composition across the whole lunar globe (Plate 2). This was followed in 1998 by another orbiter, Lunar Prospector, which provided even more insights and discoveries, such as the existence of ice dispersed within the regolith at both poles. The European Space Agency and Japan each plan lunar missions for the early years of the twenty-first century.

The atmosphere and polar ice

The Moon's atmosphere is almost as insubstantial as Mercury's, and probably has much the same origin. The Clementine mission returned our first clear views of the lunar poles, showing sites in particular near the south pole that are permanently in shadow, and which could therefore be places where ice might accumulate (Figure 6.1). Clementine's simple radar gave the first indications that water is present there, and

Name	Description	Date
Luna 2	impact with surface	Sept. 1959
Luna 3	fly-by, images of far side	Oct.1959
Rangers 7–9	images at close range prior to impact	Mar. 1965–July 1972
Luna 9	unmanned landing, pictures from surface	Feb. 1966
Luna 10	first probe to orbit Moon	Apr. 1966
Surveyor 1, 3, 5–7	unmanned landings, images from surface	June 1966–Jan. 1968
Lunar Orbiter 1–5	images from orbit	Aug. 1966–Aug. 1967
Luna 11–12	images from orbit	Aug.– Oct.1966
Apollo 8	first manned orbits	Dec. 1968
Apollo 11, 12, 14–17	manned landings, geological and geophysical studies, sample return	July 1969–Dec. 1972
Luna 16, 20, 24	unmanned sample returns	Sept. 1970–Aug. 1976
Luna 17, 21	Lunokhod unmanned surface rovers	Nov. 1970, Jan. 1973
Galileo	fly-by en route to Jupiter, compositional mapping	Dec. 1992
Clementine	high resolution multispectral imaging and laser altimetry from orbit	Jan.–Mar. 1994
Lunar Prospector	geophysical and geochemical mapping from orbit	Jan. 1998–July 1999
SMART 1	European Space Agency lunar orbiter	2003
Lunar A	Japanese lunar orbiter and seismometers	2003
Selene 1	Japanese lunar orbiter and lander	2004

table 6.1 some of the successful and anticipated missions to the Moon Luna missions were launched by the Soviet Union, the others are NASA missions unless specified

this theory was dramatically backed up by measurements by Lunar Prospector that show a reduction in the average speed of neutrons (produced by cosmic radiation) over both lunar poles. The only reasonable explanation of this seems to be collisions between neutrons and the hydrogen atoms within ice molecules, and it now seems that there may be as much as 3 billion tonnes of ice mixed with the regolith at each pole. This is not a lot in terms of the size of the Moon (it is the equivalent of a 1.5 km cube of ice at each pole), but could be ample to supply the immediate needs of human habitation on the Moon.

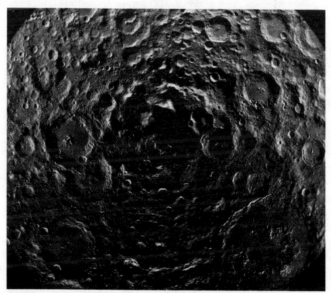

figure 6.1 a view unobtainable from the Earth, this is a 500-km-wide area centred on the lunar south pole assembled from Clementine images
much of the shadow here is permanent, and corresponds to areas where strong signs of ice have been detected
the lunar near side is towards the top, and the far side towards the bottom

The interior

The Moon is the only planetary body other than the Earth for which we have any seismic data that can tell us about its interior. Seismic stations were established at the Apollo 12, 14, 15 and 16 sites, which continued to send back data until September 1977 when operations were terminated for budgetary reasons. The Moon is seismically very quiet compared to the Earth, with moonquakes being many orders of magnitude less powerful than typical earthquakes. The Apollo seismometers also detected the vibrations from deliberate crashes of various expended units of the Apollo spacecraft and from meteorite impacts, including one from the far side that gave crucial data on the maximum possible size of the lunar core.

The picture that has emerged from combining Apollo seismic data with topographic and gravitational mapping by Clementine and Lunar Prospector is as follows. The Moon's crust is on average about 70 km thick, but exceeds 100 km in some highland regions and falls to about 20 km thick below some impact basins. The crust is thicker on the far side, with the result that the Moon's centre of mass is offset from its geometric centre by about 2 km toward the Earth.

The mantle is rigid to a depth of about 1000 km, which is also the greatest depth for the source of moonquakes, so this marks the base of the lunar lithosphere, below which it is probably slowly convecting. Like the Earth's mantle, that of the Moon is approximately peridotite in composition, though there are signs that it varies in detail from region to region.

The Moon's core is between about 220 and 450 km in radius (so the core-mantle boundary is at least 1290 km depth), and is probably iron-rich and solid. It makes up less than 4 per cent of the Moon's total mass, a much smaller fraction than for any of the other terrestrial planets. This can be explained by computer models of the giant impact in which the Moon was born. These show the core of the impacting body accreting onto the core of the proto-Earth, whereas the Moon formed from the mixture of fragments of the two bodies' mantles that were thrown into space. There is no magnetic field generated in the Moon's core today, showing that it must be solid. However, there are local weak remanent magnetic fields associated with some of the Moon's major impact basins, which suggest that part of the core may still have been liquid and generating a global field when these formed 3.6 billion years ago.

The surface

Look at the Moon even with the unaided eye, and you will see that it has dark patches on a paler background (Figure 6.2). This simple observation picks out the two distinct types of crust on the Moon. The paler areas are the **lunar highlands**, and the darker areas are the lunar 'seas' or **maria** (singular: mare). Both the highlands and the maria have very little bedrock exposed at the surface, because it is buried by several metres of regolith composed of ejecta from crater-forming impacts at all scales. In detail the regolith consists of a mixture of rock fragments, crystal fragments and droplets of glass (which are congealed

figure 6.2 1600-km-wide region of the Moon, centred on the Mare Humorum, showing clearly the distinction between the old, pale and heavily cratered lunar highlands and the darker, younger and therefore less-cratered lunar maria
note that craters are harder to make out towards the right, where the Sun was higher in the lunar sky (mosaic of Lunar Orbiter 4 images)

globules of melt produced by impacts). This forms a 'soil' capable of bearing the imprint of a booted foot (Figure 6.3, left), but there are also scattered boulders of a variety of sizes (Figure 6.3, right).

figure 6.3 left: an Apollo astronaut's footprint in the lunar regolith
right: a large fractured boulder flung out by a relatively recent impact and resting on the surface of the lunar regolith
the astronaut in the lower left provides the scale

The highlands are the first lunar crust to have formed, probably by crystals rising to the surface of the 'magma ocean' thought to have covered the young Moon as a result of the heat generated by collision of the Moon-forming fragments. Because of the preponderance of the mineral anorthite (a calcium-rich variety of the mineral plagioclase feldspar) this rock type is called anorthosite. The oldest anorthosite sample collected from the Moon has been dated at 4.5 billion years. However, most highland rock samples consist of fragments of rock that were welded together between about 4 and 3.8 billion years ago. These reflect the heavy bombardment suffered by the highland surface until the rate of impact cratering declined towards its current level.

The high degree of cratering suffered by the highland crust is notable in the view seen in Figure 6.4. This also illustrates the dependence of crater morphology on crater size. Craters less than about 15 km across are simple bowl-shaped depressions with raised rims, but the largest crater in this view has the terraced inner walls and central peak that are characteristic of craters larger than this. At yet larger diameters the central peak is

figure 6.4   a cratered region on the lunar far side, photographed from orbit by Apollo 11
the largest crater is about 80 km in diameter

replaced by a ring of peaks, as in the crater in the lower right of Figure 6.1. The largest impact structures of all are multiringed impact basins (like Mercury's Caloris basin), dating from 4 to 3.8 billion years ago. The 2500 km diameter South Pole Aitken basin (Plate 2) is the largest lunar example, but there are several others, one of which is now occupied by the Mare Humorum near the centre of Figure 6.2.

All the Moon's multiringed impact basins are older than the Moon's second kind of crust. This consists of basalt lava that has flooded low-lying areas to form the lunar maria. Many of the maria occupy multiringed impact basins, and it was once thought that the melting to produce the mare-filling basalts was a direct response to the basin-forming impacts. However, the ages of samples of mare basalt that have been returned to Earth range from 3.8 to 3.1 billion years old, so clearly the magma generation was triggered by unrelated events, and the magma ended up in the basins merely because it took advantage of the easiest routes to the surface. Inspection of detailed images of some mare areas that have not been sampled suggests that the mare-forming eruptions did not fully cease until about 1 billion years ago. The only extensive mare regions are on the lunar near side. Most of the far side features, even the South Pole Aitken basin, have escaped mare flooding, presumably because of the greater thickness of far side crust.

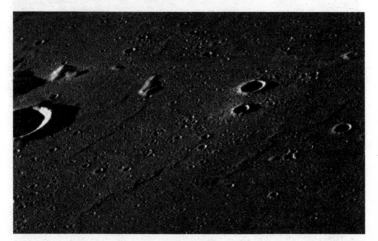

figure 6.5 8-km-wide area of the Mare Imbrium, photographed by Apollo 16 under grazing incidence solar illumination, which casts shadows at the subtle edges of some of the individual lava flows that flooded this region

The mare basalts probably came into place as a series of extensive lava flows, amounting to some hundreds of metres in total thickness. In some places, such as in Figure 6.5, the margins of individual lava flows can still be seen, and elsewhere there is spectacular evidence of the lava having flowed in tubes, which are revealed by later collapse of their roofs. A good example of this is seen in Figure 6.6, which shows Hadley Rille,

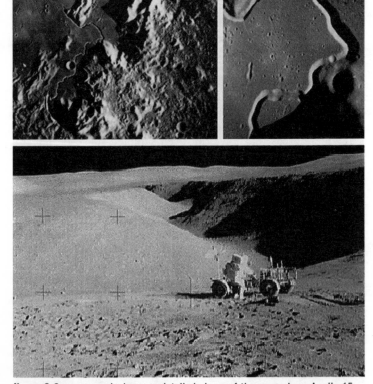

figure 6.6 progressively more detailed views of the area where Apollo 15 landed in July 1971

top left: 200-km-wide area (Apollo 15 metric camera)

top right: more detailed view of the central portion of Hadley Rille

the largest crater is 2 km in diameter, above right of it on the rim of the rille is a barely visible 300 m crater named Elbow

below: photograph looking northwest along Hadley Rille, taken from the rim of Elbow crater

the astronaut is standing beside the Lunar Roving Vehicle, which had brought him and his companion more than 3 km from the landing site the main focus of the Apollo 15 surface explorations. Most of the top left view shows the Apennine mountains, which are highland crust forming the rim of the Imbrium basin (a large near side multiringed impact basin). Hadley Rille emerges from an unseen source at the edge of the mountains, and snakes across the much flatter mare basalt terrain, appearing to have acted as a channelway that fed the later stages of mare flooding. It probably became roofed over by chilled and solidified lava, and this insulating cap enabled a vast quantity of lava to continue to flow through the tube without congealing. The rille is now visible because the roof of the tube collapsed at some time after it became drained of lava.

Living on the Moon?

The projected date for a permanent lunar base seems to have been receding into the future ever since the demise of the Apollo programme. However, unless we give up space exploration entirely it is bound to happen eventually. One major objection to lunar habitation, the lack of water, has now been removed with the discovery of substantial quantities of ice near the poles. No convincing reasons have yet emerged for basing industries on the Moon. However, apart from the advantages for lunar exploration itself, the far side would offer an excellent place from which to do radioastronomy, shielded from the increasing interference from mobile phones and the like.

07

Mars

In this chapter you will learn:
- about the Red Planet, Mars
- how its atmosphere and climate have changed over time, and about the perplexing signs that liquid water may still occasionally appear on the surface.

Planetary facts	
Equatorial radius (km)	3396
Mass (relative to Earth)	0.107
Density (g/cm³)	3.91
Surface gravity (relative to Earth)	0.38
Rotation period	24.62 hours
Axial inclination	25.2°
Distance from Sun (AU)	1.52
Orbital period	687 days
Orbital eccentricity	0.093
Composition of surface	rocky
Mean surface temperature	–50 °C
Composition of atmosphere	carbon dioxide (95%), nitrogen (2.7%), argon (1.6%), oxygen (0.13%), carbon monoxide (0.07%), water vapour (0.03%)
Atmospheric pressure at surface (relative to Earth)	0.0063
Number of satellites	2

Rotation and orbit

Mars is probably most people's favourite planet, with an exalted position in both folklore and science fiction. Furthermore, now we have begun to explore it, Mars is also turning out to be deeply fascinating scientifically.

Mars's orbit lies close outside our own. The Earth overtakes Mars about every 26 months, at which time Mars is said to be at **opposition** and reaches its highest point in our sky at midnight. At opposition, Mars can outshine even the brightest stars. However, it is not always quite so bright, because its comparatively great orbital eccentricity means that its distance from Earth at opposition varies from 56 million to 101 million km.

Mars has a rotation period not much longer than Earth's, and currently a similar axial inclination giving seasons that are Earth-like except for being nearly double the length. Seasonal changes in Mars's appearance have long been known from telescopic observation, including advances and retreats of the polar caps, obscuration by cloud of latitudes north of 40° N during the northern autumn, and global dust storms that strike when Mars is at its closest to the Sun, which currently occurs near southern midsummer.

Missions to Mars

NASA and the Soviet Union targetted Mars with space probes almost as long ago as they began to try for the Moon. The first attempts, beginning with the Soviet Union's 'Marsniks' in 1960, either failed at launch or lost contact before they reached their destination. Mariner 4 was the first successful mission, passing within ten thousand km in 1965. By ill-luck, the 1 per cent of the planet's surface that it imaged lay in a heavily cratered region, giving the impression (later reinforced by pictures sent back by the Mariner 6 and 7 fly-bys) that Mars is more Moon-like than it really is.

The first successful Mars orbiter, Mariner 9, arrived in November 1971. At first its cameras showed very little because a major dust storm was in progress. However, as the dust cleared the modern impression began to form of Mars as a world of contrasting hemispheres, and with enormous volcanoes and ancient valley systems attesting to past episodes of violent water flow, even though the atmospheric pressure is now too low to allow liquid water to survive at the surface. Under current conditions, water can exist year-round as ice at the poles or several tens of metres below ground at latitudes as low as about 40° but elsewhere the only stable form near the surface is as a vapour.

Subsequent to Mariner 9 there have been many successful and progressively more ambitious landers and orbiters, as Table 7.1 shows, and several more are planned. Looking beyond the timeframe covered by Table 7.1, NASA has plans for various small 'Scout' missions beginning in 2007, which could include aircraft or balloons. Smart landers able to detect and avoid hazards on landing may follow in about 2009, accompanied by long-duration, long-range rovers. NASA's first sample return mission is unlikely to be launched before 2014. Similarly, the European Space Agency has a strategy for future Mars orbiters and landers, leading towards sample return missions in the period 2011–2017 and perhaps a human mission about 15 years

Name	Description	Date of fly-by, landing or operational period
Mariner 4	fly-by, first close-up pictures	July 1965
Mariners 6 & 7	fly-bys, confirmed carbon dioxide in atmosphere and in polar ice	July & Aug. 1969
Mariner 9	orbiter; extensive imaging, first close-ups of Mars's satellites	Sept. 1971– Oct. 1972
Mars 2 & 3 (USSR)	orbiters and one marginally successful lander	Dec. 1971
Mars 5 (USSR)	orbiter; atmospheric measurements and some images	Feb. 1974
Vikings 1 & 2	orbiters providing extensive detailed images, and landers providing surface images and analyses	July 1976– Aug. 1980
Phobos 2 (USSR)	orbiter; failed before attempted Phobos landing	Mar. 1989
Mars Pathfinder	lander and surface rover	July–Sept. 1997
Mars Global Surveyor	orbiter; very high resolution imaging, topographic mapping	Sept. 1997– 2004
Mars Odyssey	orbiter; study of climate and geological history	2002–2004
Mars Express (Europe)	orbiter and Beagle 2 lander; stereoscopic imaging from orbit, surface studies to seek signs of life	Dec. 2003
Mars Exploration Rovers	Two rovers able to travel 100m per day	Jan. onwards 2004
Mars Reconnais- sance Orbiter	Surface studies at sub-metre resolution	March 2006
Phoenix	Polar Lander	May 2008

table 7.1 some of the successful and anticipated missions to Mars (by NASA unless specified)

later. There is no guarantee of success in every case; for example in 1999 two NASA missions (an orbiter and a lander) failed on arrival at Mars.

Russian experience in long-duration spaceflight by humans in Earth-orbit has demonstrated that astronauts could survive the long return journey time to Mars. There is no clear indication yet of when human exploration of Mars will begin, or which agency will be the first to achieve it. However, with the increasing sophistication of microtechnology, it will clearly not be long before humankind has established a routine 'virtual presence' on Mars. Billions of people will be able to participate in this using home computers linked to the worldwide web.

The atmosphere and the polar caps

Mars has a substantial atmosphere, though it contains only about a hundredth of the amount of gas found in the Earth's atmosphere. Like Venus, carbon dioxide is the most abundant atmospheric constituent, but the total density is much less and consequently the greenhouse warming of Mars's surface is only about 6 °C. The mean global temperature is about –50 °C, though noontime temperature in summer can reach above 0 °C. day–night temperature differences of more than 70 °C have been recorded at landing sites. In polar regions during winter it becomes so cold (below –120 °C) that carbon dioxide freezes out of the atmosphere, forming a frost deposit maybe a couple of metres thick that makes the polar cap appear to extend further toward lower latitudes. In spring the carbon dioxide ice begins to sublime, meaning it changes directly from solid to vapour. When this happens the polar cap shrinks, and there is a net flow of carbon dioxide through the atmosphere towards the opposite pole where the local temperature is falling to the point at which carbon dioxide begins to freeze out again.

Spectroscopic study of the residual northern polar cap in summer reveals that it consists of frozen water, showing that all the carbon dioxide frost has sublimed away. Water-ice is not visible at the surface of the residual southern polar cap, which must still retain a covering of carbon dioxide ice, but probably there is water-ice below this (Figure 7.1).

Although there is only a tiny proportion of water vapour in Mars's atmosphere, this is enough to form clouds (of ice crystals rather than water droplets), such as those picking out the spiral storm feature in Figure 7.2. Clouds can also be generated downwind of major topographic features such as volcanoes, and mists often form in low-lying areas at dawn and dusk. Frosts, presumably of water-ice, occurred during winter at the Viking 2 landing site at 47.8° N.

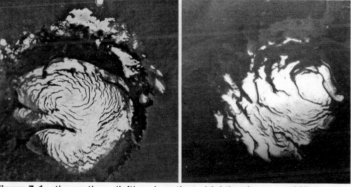

figure 7.1 the northern (left) and southern (right) polar caps of Mars during local summer when each was at its minimum extent, as seen by the Viking orbiters
each view is about 700 km across

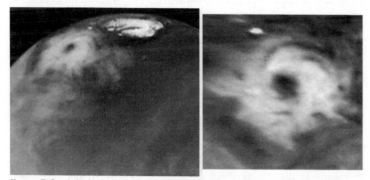

figure 7.2 left: Hubble Space Telescope image of part of the northern hemisphere of Mars in April 1999, during the northern summer
the residual north polar cap is visible at the top, and below and left of it is a spiral storm feature, picked out by clouds of condensed water vapour
less obvious, lower altitude clouds mask the surface to the south of this storm too
right: the spiral storm reprojected to show it as it would appear if seen from directly above

Martian wind, whose speed has been measured on the surface to reach as much as 30 metres per second, can pick up vast quantities of dust, especially during the retreat of the southern polar cap. This produces dust clouds that can envelop virtually the entire globe during the southern midsummer, and can take weeks to settle. Detailed images of the convoluted edges of the polar caps show that they consist of alternating ice-rich and dust-rich layers, amounting to a total thickness of a few kilometres.

Atmospheric circulation on Mars is driven by three competing phenomena. First, there is the seasonal pole-to-pole flow that has already been mentioned. Second, the strong day–night temperature change drives a flow from the comparatively warm day hemisphere (where the air expands) towards the colder night hemisphere (where the air contracts). Third, there is Hadley circulation. Hadley cells on Earth rise at the equator, because the influence of the oceans is such that the equator is the hottest latitude throughout the year. On Mars, however, the hottest latitude is where the Sun is overhead at noon. This varies with the seasons between 25.2° N and 25.2° S, and for most of the year Mars has a single dominant Hadley cell, rising at the sub-solar latitude and flowing across the equator to sink at high latitudes in the opposite hemisphere. The intensity of Hadley circulation can be temporarily increased by dust-storms, because direct heating of the atmosphere by absorption of solar radiation during dust-storms is more efficient than when heating occurs by transfer of energy from solar-heated ground under a clear atmosphere.

The interior

Mars's density is more like the Moon's than Mercury's, therefore its core must be relatively small. Current estimates for the core's probable radius range from 1300 to 2000 km. These are based on Mars's rate of precession deduced from the slight change in the position of its rotational axis over a 20-year period determined by precise tracking of the Viking and Mars Pathfinder landers as they rotated with the planet. Improved estimates must await seismic data.

Mariner 4 showed that Mars lacks a strong magnetic field, and so cannot have a liquid core. However, the 120 km close approaches to the surface by Mars Global Surveyor revealed that some areas have strong remanent magnetism, presumably frozen in from earlier times when Mars's core was still liquid and capable of generating a powerful field.

The lithosphere of Mars shows no sign of being broken into separate plates today, and may be as much as about 200 km thick. This thickness would explain its strength, which is sufficient to support the weight of the 24-km-high volcano called Olympus Mons (the largest volcano in the Solar System) without sagging. Although Mars appears to be a 'one plate planet', consideration of the global topography suggests it may have an Earth-like distinction between two types of crust.

Global topography

Mars has a 30 km range of surface height, which exceeds that of the Earth (20 km) and Venus (15 km). It is a planet of contrasting hemispheres, as can be seen in the maps of global topography in Plate 4. The southern hemisphere and part of the northern one form a vast region of heavily cratered highlands, whereas much of the northern hemisphere is low lying and occupied by sparsely cratered plains.

The relative lack of craters in the north is because of burial by lava and sedimentary material brought in from the highlands. This does not demonstrate that the basement to this vast northern hemisphere basin, which is buried, has to be younger than the southern highlands.

Whatever the means by which the northern lowlands were created, the underlying crust there is likely to be thinner, and probably younger, than that of the southern highlands. The present-day topographic boundary between the two regions consists of a convoluted scarp whose shape and location have clearly been modified by erosional retreat (Figure 7.3). Presumably this is now some way south of the actual junction between the two types of crust.

The most ancient region of the southern highlands shows hints of magnetic stripes similar to those mapped on the Earth's ocean floors. These may indicate that the crust here grew by spreading apart at a constructive plate boundary, in which case the stripes would indicate that Mars's magnetic field reversed direction several times during the process. The more recently disturbed areas of the highlands lack this magnetic signature, possibly because the planet had ceased to generate its own magnetic field by then.

The picture of the simple high-low, south-north distinction on Mars, sometimes referred to as the 'crustal dichotomy', is complicated by three factors. The first is the occurrence of a number of impact basins. The largest of these, Hellas, is a deep

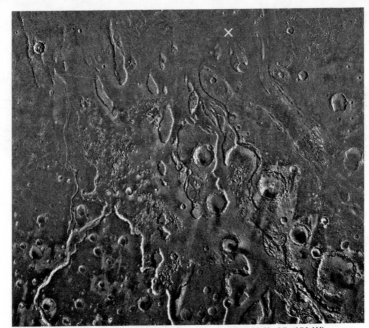

figure 7.3 image-based map of part of Mars (3°–21° N, 25–45° W)
showing the transition between the heavily cratered southern highlands and
the low-lying northern plains
channels from the highlands debouch into the plains, where several
erosional remnants of highland terrain survive as streamlined 'islands'
beyond the mouths of these channels
the site where Mars Pathfinder landed in 1997 is indicated by a cross
the Viking 1 lander touched down just beyond the northwestern corner of
this view in 1973
the area shown is about 1000 km across and illumination is from the right

2000 km diameter basin within the southern highlands that is
obvious in the upper right view in Plate 4.

The second is a 4000-km-wide 'bulge' straddling the north-
south boundary in the Tharsis region, where both types of crust
are uplifted. Several of Mars's most recent volcanoes occur on
the Tharsis bulge, which is in the western part of the lower left
of the four global views in Plate 4. The youngest and biggest
volcano, Olympus Mons, lies on its western flank. The bulge is
generally reckoned to owe its origin to thermal uplift of the
lithosphere over an exceptionally persistent hot upwelling zone

in the mantle (a mantle plume) that fed the local volcanoes. Gravity data suggest that the thermal support has gone now, and that the bulge survives because the lithosphere has become too strong and rigid to allow it to subside again.

The third complicating factor is a giant complex of canyons, named Valles Marineris (after Mariner 9, which discovered them), on the eastern side of the Tharsis bulge. They may have been originated by fracturing of the crust over the arch of the bulge, although flowing water evidently deepened these canyons and landslips widened them into the impressive spectacle we see today, which would dwarf the Earth's Grand Canyon. Valles Marineris can be seen in the eastern part of the lower left globe in Plate 4, and dominates the simulated orbital view at the bottom of the plate.

The surface

The surface of Mars is famous for its redness. This is apparent at all scales from the global colour of Mars as seen in our sky down to close-up views obtained by landers (Plate 5). The reason is that Mars's surface is a highly oxidizing environment, and the colour we see is essentially that of rust. The conditions are so oxidizing not simply because of the oxygen in the atmosphere but because of minute traces of highly reactive compounds such as ozone (O_3) and hydroperoxyl (HO_2), which are produced by the action of solar ultraviolet radiation on oxygen and water vapour. Thus on rock surfaces the iron in silicate minerals is oxidized, and these minerals eventually break down to a red dust consisting of hydrated iron oxides (rust) and iron-rich clay.

Landing sites

The Sojourner rover deployed from Mars Pathfinder in 1997 analysed the composition of several rocks, and found that the closest match to terrestrial rocks is with andesites. However, interpretation of the data is open to doubt because the analyses were made through a veneer of dust. The close-up pictures of the rocks show a variety of textures, indicating that some rocks may be volcanic whereas others are sedimentary, consisting of fragments held together in a cemented sandy matrix. There are also a few meteorites collected on Earth that are believed to be samples of Mars rock that were flung into space as fragments of ejecta from major impact craters. These are igneous rocks of generally basaltic composition, and the youngest is only 200 million years old indicating that igneous activity on Mars has continued into comparatively recent times. Thus, all the signs are that Mars has had a long and complex history of rock formation and redistribution.

The two Viking landers and Mars Pathfinder all landed north of the dichotomy boundary, principally to take advantage of the lower altitude, which allowed more effective use of parachutes to slow the rate of descent for sufficiently soft landings. At all three sites the landers' cameras revealed rock-strewn landscapes (Figure 7.4, Plate 5). Viking 1 and Mars Pathfinder landed in the 'outwash' areas of channels draining from the highlands, and many of the rocks there must have been transported into their present positions by flooding. However, some of them, and perhaps most of the rocks at the Viking 2 site, may be ejecta from impact craters.

figure 7.4 the Mars Pathfinder lander as seen from 4 m away by the Sojourner rover in 1997
note the irregularly shaped rocks of all sizes resting on top of the soil

Craters

Impact craters are common all over Mars, but especially in the southern highlands where their abundance suggests a surface age of about 3.8 billion years. The crater density on the northern lowlands varies from place to place, showing that the lowland surfaces were deposited over a wide time interval, perhaps continuing as recently as 1 billion years ago. Some volcanic regions are even younger. The older craters have been noticeably degraded by erosion or infill by sediments, and

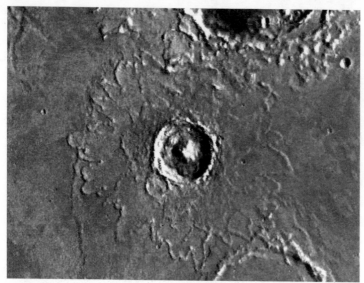

figure 7.5 a lobate ejecta blanket around the crater Yuty, which is about 20 km across

examples of both kinds can be seen in Figure 7.3. However, younger craters retain their original form, which is notable in many cases for the lobate nature of their surrounding ejecta blanket (Figure 7.5). This may indicate that at the time of impact the ground contained either ice (like arctic permafrost on Earth) or water, so that some of the ejecta flowed as a slurry. Alternatively it could simply be because the atmosphere did not allow the ejecta to disperse so freely as it would have done in a vacuum (as on the Moon or Mercury, for example).

Channels and gullies

The southern highlands are cut by a variety of channels (as distinct from the Valles Marineris canyon complex). Typically, these are steep sided and usually meandering valleys. Many of the older ones have branching systems of tributaries, and have the appearance of drainage systems fed by rainfall. Others (of a wide range of ages according to the cratering timescale) lack well-developed tributary systems and are more likely to have been supplied by catastrophic escape of previously frozen groundwater, perhaps triggered by an impact or volcanic eruption. The channels in Figure 7.6 and also in the southern

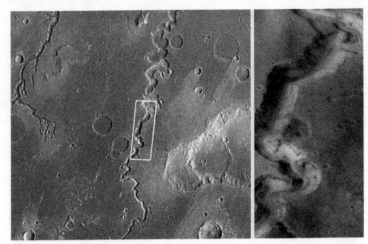

figure 7.6 left: Viking Orbiter view showing two channel systems in Mars's
southern highlands
the area shown is 110 km across, and the white outline indicates the area
shown in more detail on the right
right: high resolution Mars Global Surveyor view of part of one of these
channels, whose walls show horizontal stripes, which could either be
terraces indicating multi-stage evolution of the valley, or reflect layers in the
bedrock through which this channel has been carved

part of Figure 7.3 are of this type. The peak rate of discharge
along these must have been enormous, and led to the sculpting
of streamlined 'islands' on the edge of the northern lowlands
into which they discharged, distributing sediment in the form of
lumps of rock and sand across the plains. Some channel systems
may have held flowing water for the duration of just one major
flood event, others show hints of multi-stage flow.

Some of the very detailed images obtained by Mars Global
Surveyor (Table 7.1) surprised everybody by revealing gullies
only a few tens of metres wide descending steep slopes on the
walls of canyons and craters (Figure 7.7). It is hard to imagine
anything other than liquid water having cut these gullies. Many
of the gullies have an embayment or 'alcove' at their uphill end,
which looks as if it is a result of undermining by water escaping
from the subsoil rather than being cut by surface run-off as a
result of rainfall. The pristine shapes of the gullies indicate that
they must be very young features, and so does the observation

that the fan of debris beyond the downhill end of each gully often overlies other clearly young features such as sand-dunes (as in the lower right of Figure 7.7).

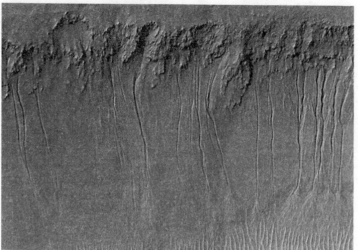

figure 7.7 a 2-km-wide Mars Global Surveyor image showing young, presumably water-cut, gullies on the wall of a larger channel

Wind effects

Although water flow on Mars has been both episodic and ephemeral, wind continues to play a role in sculpting the planet's surface. This occurs both through redistribution of sediment such as dust in dust-storms and sand to form sand-dunes (Figure 7.8), but also through erosion, especially of isolated hills which can develop some bizarre wind-sculpted shapes (Figure 7.9).

Volcanic activity

Volcanoes occur notably in the Tharsis province, but also in several other areas of both hemispheres. Many of them, including the Tharsis volcanoes which are the youngest, have the form of basaltic shield volcanoes similar to those known on the Earth and Venus, except that Mars's shield volcanoes are considerably bigger. The largest, Olympus Mons (Figure 7.10), has about 100 times the volume of its largest equivalent on Earth, which is the island of Hawaii. Such an enormous volcano is

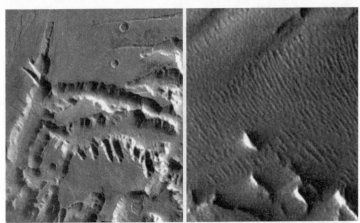

figure 7.8 left: 250-km-wide Viking Orbiter view of part of Noctis Labyrinthus, the highly fractured terrain at the western end of the Valles Marineris canyon system
right: highly detailed view of part of the canyon floor only 1.5 km across from the centre of the view on the left, as seen by Mars Global Surveyor, revealing a swarm of 50-m-wide sand-dunes oriented northwest to southeast which indicates either a southwesterly or a northeasterly prevailing wind

there are some larger hillocks sticking up through the sand, which could be blocks that slid off the canyon walls during landslides

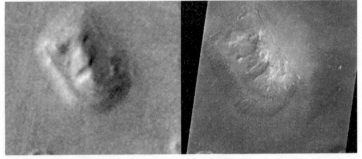

figure 7.9 a 5-km-wide area just north of Mars's crustal dichotomy containing an eroded remnant of highland terrain surrounded by plains material, as seen by Viking Orbiter under evening illumination with a low Sun (left) and Mars Global Surveyor under morning illumination with a high Sun (right)
in the view on the left the hill resembles a human face, but the likeness vanishes in the differently illuminated view on the right, which has ten times higher resolution

presumed to be possible only because Mars's lithosphere is thick and strong, and has remained stationary with respect to the underlying mantle plume that has supplied magma to the volcano. It is impossible to put a precise date on Olympus Mons. Despite its size it is so young that too few impact craters occur on it for us to put much trust in the cratering timescale. However, its last previous eruption may have been as recently as a few tens of millions of years ago. If that is the case, then there is no good reason to suppose that it has turned off for good.

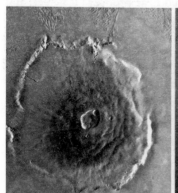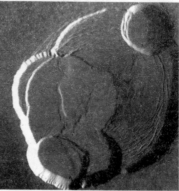

figure 7.10 left: 800-km-wide view showing Olympus Mons, which rises 24 km above its base and is the largest volcano in the Solar System
right: enlarged view of the 90-km-wide complex of overlapping calderas at the summit (illumination from the right)
these calderas are volcanic craters, formed by subsidence in response to the eruption of lava

Global history and climate change

The global history of Mars is currently understood in a general sense only. The sequence of events can be deduced, but estimates of the dates at which they occurred could be wildly inaccurate.

Broadly speaking, we can imagine highland crust forming about 4.5 billion years ago, initially in a similar manner to the Moon's, but later continually modified by intrusive and volcanic igneous activity. At some ill-defined date, but almost certainly longer ago than 3 billion years, the northern lowlands were formed by an unknown process. At about the same time, or later, the Tharsis bulge began to develop, and the Valles Marineris canyon system began to form as a series of fractures, later deepened by flowing water and widened by landslips. Meanwhile, volcanic activity took place in diverse regions of the planet, sometimes triggering the release of catastrophic quantities of flood water from imprisonment as ice in the subsoil. There may occasionally have been lakes or even a shallow sea in the northern hemisphere. Gradually, Mars lost more and more of its atmosphere to space, so that its climate became progressively more sterile, and the frequency and intensity of volcanic activity declined as Mars's rate of internal heat production waned and its lithosphere thickened.

There were probably short-term climatic fluctuations superimposed on the gradual trend towards more hostile surface conditions. These would have been caused by cyclic variations in the amount of tilt of Mars's axis, which under tidal influence from Jupiter wanders between 5° and 35° on timescales of between 100,000 and 10 million years. When the tilt is least, we would expect summertime ice loss from the polar caps to be least, and when the tilt is greatest there should be more loss from the summer pole and more accumulation at the winter pole. This could explain the alternation of dust-rich and ice-rich layers revealed around the edges of the polar caps.

The age of the youngest water-cut gullies, such as those in Figure 7.7, is controversial. Estimates range from ten million years to as little as ten years. Perhaps water escapes from the frozen subsoil to cut these gullies only during exceptionally warm parts of climate cycles, so that gullies all over the globe tend to be active at roughly the same time. On the other hand maybe melting of ice in the subsoil is triggered by local events such as landslips or shallow intrusion of molten rock, in which case gulley-forming water-escapes could happen at any time.

Life on Mars

Clearly the fantasy of a native Martian civilization surviving today is unsustainable in view of what we know of the current conditions on Mars. However, the evidence for a denser, wetter and (because of the greenhouse effect) warmer atmosphere in Mars's distant past is compelling. The environment then would have been more amenable to familiar forms of life, so if life could have started on Mars in the same way that it appeared on

Earth there could have at least been flourishing plant life equivalent to terrestrial lichens and mosses at the times when the large channels were in flood. Such organisms probably could not tolerate today's conditions on Mars. However, life as a whole is very tenacious and varieties of microbes (bacteria and the like) have been discovered on Earth surviving in conditions that are extreme by 'normal' standards. Of most relevance for Mars are photosynthetic organisms that live within Antarctic ice, algae living beneath rock surfaces and feeding on chemical energy or sunlight that penetrates transparent crystals, microbes surviving in salt deposits a kilometre below the surface, and others living in volcanic hot springs.

Debate is likely to continue on how to interpret apparent chemical and isotopic traces of life in the Martian meteorite known as ALH84001, and what look like exceptionally small fossilized bacteria within it. Some experts claim this as evidence of life of Mars, others say there are non-biological explanations, and others explain these features as contamination that occurred after the meteorite arrived on Earth. For a claim that life exists (or has existed) on Mars to be accepted, it probably has to be based on analyses performed either actually on Mars or on samples returned from Mars under controlled conditions.

The two Viking landers scooped up some soil and performed a few simple tests to look for life, including adding radioactively tagged nutrients. However, although some radioactive carbon dioxide was produced this is now understood to have been as a result of a chemical reaction with the highly oxidizing soil rather than a product of respiration by living organisms.

To find life on Mars we have to look in the right places, and the exceptionally hostile topsoil no longer seems a likely setting. The total biomass of life on Mars, if it exists, is likely to be proportionally far less than on Earth, so to find signs of life (either living or fossil) it will be important to use well-chosen techniques. If all goes to plan (Table 7.1), early in 2004 the Beagle 2 lander will grind up samples of rock and soil taken from below the protection of a boulder and make sophisticated analyses of carbon isotopes, which are good indicators of biological processes. If this strategy proves inconclusive, the question of whether there is now, or ever has been, life on Mars will presumably remain unanswered until larger samples are returned to Earth for fuller analysis.

Satellites

Name	Distance from planet's centre (km)	Radius (km)	Orbital period	Mass	Density (g/cm³)
Phobos	9377	13.1 × 9.3	7.65 hours	1.08×10^{16} kg	1.9
Deimos	23463	7.8 × 5.1	30.3 hours	1.8×10^{15} kg	1.8

table 7.2   satellites of Mars

All the planets beyond the Earth have satellites. Those of Mars, named Phobos and Deimos, are small and are simply asteroids that have been captured into orbit about the planet. Some of their properties are listed in Table 7.2. They are not long-term companions of Mars, perhaps having been captured only within the past billion years. Phobos is in an especially low orbit, less than 6000 km above the planet's surface, and tidal interactions are forcing this orbit to decay to the extent that Phobos is likely to strike the surface of the planet within about 50 million years.

Images of Phobos and Deimos are shown at their correct relative scales in Figure 7.11. Both have the irregular shapes expected of collisionally shaped objects whose self-gravity is far too slight to pull themselves into the spheroidal shape that is assumed by larger planetary bodies.

Phobos is notable in being crossed by arrays of grooves. It has been suggested that these represent fractures caused by the impact that generated the large crater Stickney. However, given that most grooves appear to be strings of overlapping craters, an attractive alternative explanation is that they are 'machine-gun bullet' trails caused by Phobos passing through streams of ejecta emanating from large crater-forming impacts on the surface of Mars.

Because Phobos and Deimos are captured asteroids, discussion of their compositions is deferred until the next chapter, which deals with asteroids in general.

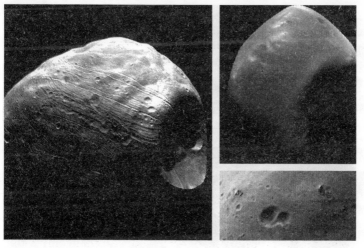

figure 7.11 Phobos (left) and Deimos (right) at the correct relative scales, as seen by the Viking orbiters
note the shadows in hollows (mostly impact craters) near the day–night boundary on each
Deimos appears less cratered only because the resolution of the original image is poorer than for Phobos
the largest crater on Phobos, named Stickney, is 10 km across
the image on the lower right is 1.4-km-wide detail from a Mars Global Surveyor view of the surface just beyond Stickney's outer edge, obtained from a range of about 1080 km and showing craters down to the limit of resolution and several boulders scattered across the surface

08

asteroids

In this chapter you will learn:
- about the rapidly expanding field of asteroid research
- about typical and unusual orbits, the variety of asteroid compositions, and the risks of asteroid impacts onto the Earth.

Tens of thousands of asteroids have been discovered, and there are probably far more awaiting detection that are less than a few km across. Known asteroids range in size from 933 km across in the case of the largest, Ceres (which was the first to be discovered, in 1801), to bodies less than 10 m in size (that can be detected only if they pass close to the Earth). Presumably the size spectrum continues down to smaller pieces of the size of common meteorites. Most asteroids are elongate and irregular in shape, many of them even more so than Phobos, and some have small satellites.

Rotations and orbits

The properties of a few exemplary asteroids are listed in Table 8.1. 'Typical' (or 'main belt') asteroids have orbits that keep them somewhere between Mars and Jupiter. This is the case for all those listed in the table except for Agamemnon, Chiron, Apollo, Eros and Braille. Jupiter's gravitational influence on the asteroids has been not just to inhibit growth by collision and to fling many of the asteroids out of the Solar System (as described in Chapter 02), but also to affect the orbits of most of the surviving asteroids. Virtually no asteroids have orbital periods

Name	Mean distance from Sun (AU)	Orbital eccentricity	Radius (km)	Orbital period (years)	Rotation period (hours)
Ceres	2.768	0.097	467	4.60	9.08
Pallas	2.771	0.180	261	4.61	7.81
Vesta	2.361	0.090	255	3.63	5.34
Agamemnon	5.233	0.02	83	11.97	7?
Chiron	13.65	0.067	100	50.4	5.92
Apollo	1.471	0.56	0.8	1.78	3.06
Mathilde	2.646	0.27	29 × 25	4.30	418
Ida	2.862	0.04	28 × 7.5	2.862	4.63
Eros	1.458	0.22	16.5 × 6.5	1.76	5.27
Gaspra	2.209	0.17	9 × 4.5	3.28	7.04
Braille	2.341	0.43	1.1 × 0.5	3.58	?

table 8.1 characteristics of the three largest asteroids, a Trojan asteroid (Agamemnon), the largest of the Centaur class (Chiron), an Earth-crossing asteroid (Apollo), and those asteroids visited by spacecraft (? = exact data unknown)

that are simple 4:1, 3:1, 5:2 or 2:1 ratios of Jupiter's orbital period, where they would be in **orbital resonance** with Jupiter. These unoccupied parts of the asteroid belt are known as the Kirkwood gaps. However, there is a fair sized family of asteroids whose orbital periods are two-thirds that of Jupiter (a 3:2 orbital resonance).

Some 1600 known asteroids actually orbit at the same average distance from the Sun as Jupiter, and therefore have the same orbital period. These (e.g. Agamemnon in Table 8.1) are known as the Trojans. Although the Trojan asteroids share Jupiter's orbit, they can be found only close to two specific points in space, 60° ahead of and 60° behind the giant planet. These two **Trojan points** are equidistant from the Sun and Jupiter and are gravitationally stable locations, about one or other of which the position of each Trojan asteroid oscillates.

A few asteroids have been discovered occupying the Trojan points of Mars's orbit. The largest is called Eureka and is only about 2 km across. Despite detailed searches, no asteroid has yet been found at either of the Earth's Trojan points. However there are four known to share the Earth's orbit in a more complex way. The largest of these is the 5 km diameter Cruithne (a Gaelic name pronounced 'Croo-EEN-ya'). This completes one orbit of the Sun every year, but has an eccentric orbit that takes it out almost to the orbit of Mars and in almost to the orbit of Mercury. Relative to the moving Earth, this path has the shape of a giant kidney bean that migrates slowly along the track of the Earth's orbit without ever coming within 0.1 AU of the Earth.

There are also some asteroids called Centaurs orbiting beyond Saturn (e.g. Chiron in Table 8.1). Continuing discoveries of bodies in this region suggests that there is probably no clear-cut distinction between asteroids, comet nuclei and Kuiper belt objects.

All known asteroids orbit the Sun in the prograde direction, though it is not uncommon for their orbits to be inclined to the ecliptic by up to about 40°. Orbital eccentricities are typically greater than for the planets. Some, such as Eros and Braille, pass inside the orbit of Mars whereas some, such as Apollo and Cruithne, cross the Earth's orbit. Two are known whose perihelions are closer to the Sun than Mercury. Orbits of those that come close to planets are unstable, because they are perturbed by the planet's gravity, and such an asteroid will typically survive only a few million years before either striking the planet or being flung out of the Solar System.

Typical rotation periods, which in most cases are deduced from periodic variations in apparent brightness as seen through a telescope, are about nine hours. The fastest known is less than three hours and the slowest is about 50 days.

Asteroid names

Most of the asteroids referred to so far have names that you might recognize from mythology. These tend to be the larger asteroids, which were discovered longest ago. Names are no longer restricted to mythological characters, but must be inoffensive and not connected with recent political or military activity. However, no asteroid is awarded a name until it has been observed long enough for its orbit to be determined with a fair degree of precision. This may take several years, but when it is achieved the body is awarded a 'permanent designation' (a number usually issued in strict numerical sequence) and the discoverer is invited to suggest a name for approval by a special committee of the International Astronomical Union. Strictly speaking, Ceres, the first known asteroid, is known as (1) Ceres, Eros as (433) Eros, and Chiron as (2060) Chiron, but in this book you will usually find only the name. Examples of more unusual or whimsical names include Poulanderson (named after a science fiction author), Beatles (named after the 1960s' pop group), and Tsenaat'a'i (which means 'flying rock' in the Navaho language). The prize for the most imaginative name goes to (2037) Tripaxeptalis. The name is pure invention but sounds like 'triPax-septAlice', which reflects the fact that its permanent designation is three times that of (679) Pax and seven times that of (291) Alice.

Until its orbit has been sufficiently well-documented, each new discovery is known only by a 'provisional designation' consisting of the year of discovery followed by two letters and, if necessary, numbers that relate more precisely to the date and sequence of discovery. For example, the asteroid Braille, listed in Table 8.1, initially had the provisional designation 1992 KD. The K signifies that it was discovered during the period 16–31 May and the D shows that it was the fourth discovery during that period. The letter I is not used in this convention, so the remaining 25 letters of the alphabet enable 25 asteroids to be designated during the period. When this number is exceeded, the letter code sequence is repeated as many times as necessary with a numerical subscript that is incremented every 25 discoveries. For example, in the designation of 1998 SF_{36}, a small asteroid that you will meet shortly, the S signifies that it was discovered during the period September 16–30, and the F_{36} is code indicating that it was the 906th body to be discovered in that period. You can tell from this example that modern telescopic searches have led to discovery of small bodies at an amazing rate, though in fact the provisional designation codes are shared with other discoveries such as Kuiper belt objects so not all of them are asteroids.

Missions to asteroids

Name	Description	Date of fly-by landing, or operational period
Galileo	fly-bys of Gaspra and Ida, en route to Jupiter	Oct. 1991, Aug. 1993
NEAR (Near Earth Asteroid Rendezvous)	fly-bys of Mathilde and Eros, orbital study of Eros	June 1997, Dec. 1998, Feb. 2001
Deep Space 1	fly-by of Braille; obtained spectra but imaging cameras misdirected	July 1999
Muses-C (Hayabusa)	sample return from Itokawa	arrival Oct. 2005, sample return June 2007
Rosetta	fly-bys of asteroids, en route to comet rendezvous	launch 2004, arrival 2014
Dawn	orbiter at Vesta (2010–11) and Ceres (2014–15)	launch 2006

table 8.2 some of the successful and anticipated missions to asteroids these are NASA, except for Rosetta (European Space Agency), and Muses-C (ISAS, Japan)

Table 8.2 lists previous and anticipated missions to asteroids. Two spaceprobes have made successful encounters with asteroids, other than Phobos and Deimos, which now orbit Mars (Chapter 07). The first was the Jupiter-bound probe Galileo, which flew past Gaspra in October 1991 and Ida in August 1993. On the latter occasion it made the remarkable discovery that Ida has a $1.6 \times 1.4 \times 1.2$ km diameter companion in orbit about it, which has been named Dactyl (Plate 6).

The second successful mission was NEAR (Near Earth Asteroid Rendezvous). This flew past the largest asteroid yet visited, Mathilde, in June 1997 at a range of 1200 km on its way to a rendezvous with the smaller asteroid Eros in December 1998. It was supposed to go into orbit about Eros for a protracted period of detailed study, but a crucial engine burn failed to happen, which delayed the start of this important phase of the mission until the next encounter in February 2000. This time it successfully entered orbit about Eros, from where it studied it in detail before gently landing on its surface a year later. Asteroid images from Galileo and NEAR are shown in Figures 8.1 and 8.2.

figure 8.1 the first four asteroids to be visited by spacecraft, shown at their correct relative sizes
clockwise from left: Mathilde, Gaspra, Ida (differently oriented than in Plate 6), and Eros

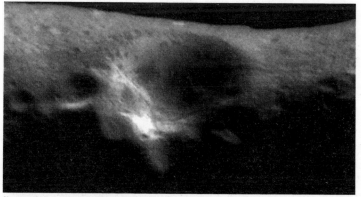

figure 8.2 18-km-wide close-up of Eros seen from orbit by NEAR
note the potential grooves cutting the right-hand wall of the largest crater, and a 50-m-wide boulder on its floor (compare with Phobos in Figure 7.11)

A third spacecraft, Deep Space 1, made a partially successful fly-past of a 2.2 × 1 km asteroid Braille in July 1999. It was able to determine Braille's surface composition spectroscopically, but the opportunity for close-up imaging was lost because of a camera pointing error.

Future asteroid missions are planned by several of the smaller space agencies. The Japanese Space Agency's Muses-C is scheduled to arrive at a 400-metre diameter asteroid (25143) Itokawa in September 2005 and return a sample to Earth in June 2007. In roughly the same period, the European Space Agency's Rosetta probe will make fly-bys of two asteroids on its way to a comet rendezvous (Table 8.2). There is also a US-based commercial venture planned under the name of NEAP (Near Earth Asteroid Prospector) that may be able to achieve a landing on an asteroid before 2010.

Asteroid compositions

So far as can be judged from their colours and other spectral properties, asteroids show the full range in compositions exhibited by meteorites such as iron, stony iron and various kinds of stony material, including carbonaceous chondrites. Although there is a lot of scatter, there is a general correlation of colour-type with distance from the Sun.

Conventional stony asteroids make up the majority of the population between about 2 and 2.6 AU from the Sun, but from 2.6 to 3.4 AU apparently carbonaceous asteroids are the most common. This is where asteroids akin to the iron meteorites are at their most abundant too. Beyond 3.4 AU, most asteroids are dark and red in colour. This type is apparently unrepresented by meteorites, and the colour is probably caused by tarry substances called **tholins** that can form from carbonaceous material or methane ice during prolonged exposure to solar radiation, a process described as 'space weathering'. The third-largest asteroid, Vesta, appears to have expanses of basalt on its surface, indicating that its interior was once hot enough for melting, which is surprising in so small a rocky body.

Of the asteroids that have been visited by spacecraft, Phobos and Deimos probably belong (surprisingly) to the dark red tholin-rich group; Gaspra, Ida and Eros are of the stony type; Mathilde has a carbonaceous appearance; and Braille is basaltic and could be a fragment from Vesta. 1998 SF_{36}, the target asteroid for the Muses-C mission, appears similar to stony meteorites.

Perturbations of Mars-orbiting spacecraft by Phobos and Deimos show that their densities are about 1.9 and 1.8 g/cm^3 respectively, and the deflection to NEAR's trajectory by Mathilde indicated a density of about 1.3 g/cm^3. Galileo's trajectory was not sufficiently deflected by either Gaspra or Ida to use this method, but the orbital characteristics of Ida's satellite Dactyl suggest a density in the range 2.0–3.1 g/cm^3. A better determined orbit of another asteroidal satellite, a 9 km radius satellite (discovered telescopically in 1998 and named Petit-Prince) of the 107 km radius carbonaceous asteroid Eugenia, gives Eugenia a bulk density of 1.2 g/cm^3. The best determination of all comes from NEAR's orbit about Eros, which reveals a density of 2.7 g/cm^3.

In contrast, stony-iron meteorites and stony meteorites have densities of about 5 and 3.5 g/cm^3 respectively, so these density determinations reveal that asteroids in general are not solid bodies. Rather they must contain ice or internal voids, which could be spaces from which volatile material, such as ice, has been lost. Probably, many asteroids are heaps of regolith-covered rubble. This would fit with their low densities and also the elongated, and in some cases narrow-waisted, shapes of many of them (notably Ida, Eros and Gaspra in Figure 8.1), which could have been caused by tidal stretching of loosely bound rubble during a close encounter with one of the planets but would be harder to explain as a result of collision. The most elongated asteroid known is Geographos, whose dimensions were measured as 5.1 km × 1.8 km by radar when it passed by the Earth in 1994.

Asteroids with satellites

Technological advances in ground-based telescopes, notably a technique known as adaptive optics that is able to compensate for the shimmering of the Earth's atmosphere, have revealed satellite companions to several asteroids. Petit-Prince, the tiny satellite of Eugenia, was the first to be detected in this way. Examples now known include other large asteroids with small satellites (such as the 250-km-diameter Camilla with a 10-km-diameter satellite), large double objects (such as Antiope, which is revealed as two 80 km objects about 160 km apart), and small double objects (such as 2002 KK_8, which consists of a closely-spaced pair about 500 m and 100 m across). One of Jupiter's Trojan asteroids, Patroclus, that was previously thought to be a single body about 140 km across has also been shown to consist of a pair of bodies about 100 km and 90 km across. Bearing in mind how common asteroid satellites and double asteroids are turning out to be, it seems likely that they represent fragments created by collisions when the asteroids were being formed (see Chapter 02).

Surface features

Close-up images of asteroids reveal craters at all sizes, but the abundances of craters of different sizes varies from body to body. For example Gaspra has fewer large craters relative to its size than do either Ida or Mathilde. Mathilde is not elongated, instead its shape is dominated by at least four craters whose diameters are greater than the average radius of the body itself. Mathilde's unusually slow 17-day rotation period may be because one or more of the large crater-forming impactors hit the surface from precisely the right direction for its momentum to cancel out Mathilde's previously faster rotation.

Ida, Gaspra and Eros all have grooves on their surfaces, but not so common or so well-defined as those on Phobos. They could be fractures induced by impact or the hypothetical tidal stretching responsible for their shapes, or may represent fissures from which trapped volatiles escaped.

A threat to the Earth?

Estimates suggest that there may be as many as a thousand 'Earth-crossing' asteroids whose orbits bring them inside the Earth's that are 1 km or greater in diameter, and untold numbers of related objects smaller than this. They appear to have been ejected out of the 3:1 Kirkwood gap. As noted above, theory suggests that their current orbits are transient, and that these bodies are likely to crash into the Sun or a terrestrial planet, or else be scattered out of the Solar System, within a few million years.

Statistical arguments show that about five asteroids 1 km or greater in size are likely to strike the Earth every million years on average. Undoubtedly such objects have struck the Earth many times. The largest impact for which we have direct evidence happened 65 million years ago, and was associated with a mass extinction event that wiped out the last of the dinosaurs and many other kinds of life on land and in the oceans. This impactor must have been either an Earth-crossing asteroid or a comet about 10 km across.

Events of this magnitude are fortunately very rare, but impact by any asteroid big enough to penetrate the atmosphere with unabated speed would have globally significant consequences. A stony object less than about 100 m across that strikes the Earth's atmosphere is likely to become sheared apart into pieces small enough to be slowed down so much by aerodynamic drag that they would not hit the ground at crater-forming speed. However, objects bigger than this must be considered as potential major hazards. If one struck the sea the resulting 'tidal wave' (more properly described as a tsunami) would break across the surrounding coastlands to a considerable height. If the impact was on land, apart from the total destruction within the crater, the accompanying blast wave would devastate a wide surrounding area. The loss of life and economic dislocation caused by the immediate damage in either case would be enormous, and to make matters worse enough debris could be thrown into the atmosphere to block out enough sunlight to cause global famine.

It has been calculated that, averaged out over millions of years, the chance of any individual being killed by an impact is comparable with the chance of dying as a result of an aircraft accident and perhaps 30 times less likely than death by accidental electrocution. These are not insignificant odds, and so efforts have begun (despite a lack of co-ordinated central funding) to detect and catalogue all the 1-km-diameter objects in near-Earth orbits ('near-Earth asteroids') to see if any of them is likely to strike the Earth in the foreseeable future, and to give as much warning as possible of the approach of smaller objects which cannot be detected until they get close. So far, the closest approach observed to Earth was by a 30-m-diameter object designated 2004 FH that came within 43,000 km in March 2004. The closest predicted approaches for known bodies before the year 2050 are of a 600-m-diameter asteroid 2001 WN_5 that will pass us at a range of 255,000 km on 26 June 2036, and the 200 m asteroid 2003 MK_4 that will pass at only 210,000 km on 3 January 2032. Fortunately for us, these are both actually comfortable misses. After discovery of an Earth-crossing asteroid heading towards the Earth there is usually a period of several weeks before its orbit is known with sufficient precision to rule out a collision. It is this period of uncertainty that gives rise to such news headlines as 'astronomers discover asteroid on collision course with Earth'.

It is difficult to see what we could do about it were we to discover even a small asteroid above the 100 m size limit on a collision course. If small asteroids are rubble piles like their larger cousins such as Mathilde, Ida and Eros, then attempts to deflect them with nuclear devices might simply disrupt them into several fragments. These might cause even more widespread destruction, unless all the fragments were smaller than 100 m. The deflection of double objects would be especially problematic.

On a more positive note, non-colliding near-Earth asteroids could prove to be of major economic benefit during industrial development of near-Earth space. Iron and stony asteroids could prove valuable not only for their primary metals such as iron and nickel but also because of their likely richness in platinum and related elements. Carbonaceous asteroids could turn out to be convenient sources of rocket fuel. The era of commercial investment in asteroid resources is now dawning, and it is not beyond the bounds of possibility that somebody will be making a profit before 2020.

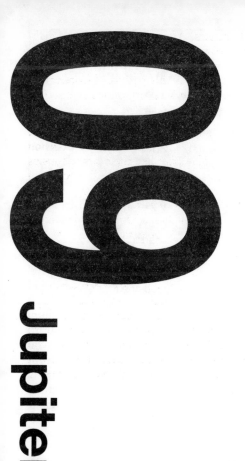

09 Jupiter

In this chapter you will learn:
- about the largest giant planet in the Solar System: about its dense interior, its deep and dynamic atmosphere, and what happened when it was struck by comet fragments in 1993
- about Jupiter's slender rings and amazingly diverse family of satellites.

Planetary facts	
Equatorial radius (km)	71,492
Mass (relative to Earth)	317.7
Density (g/cm³)	1.33
Surface gravity (relative to Earth)	2.36
Rotation period	9.93 hours
Axial inclination	3.1°
Distance from Sun (AU)	5.20
Orbital period	11.86 years
Orbital eccentricity	0.048
Composition of surface	gassy
Mean cloud-top temperature	−150 °C
Composition of atmosphere	hydrogen (90%), helium (10%), methane (0.3%), ammonia (0.03%)
Number of satellites	61

At Jupiter, the character of the Solar System takes a new turn. Jupiter is the first of the giant planets, and contains more mass than all other planetary bodies in the Solar System put together. Its internal structure is compared with that of the other giant planets in Figure 2.5.

Rotation and orbit

Jupiter's orbital velocity is about 13 km per second, little more than half that of Mars, allowing Earth to overtake it every 399 days. Jupiter's orbit is less eccentric than Mars's, and its distance at opposition varies between 590 million and 670 million km. Its large size and the reflectivity of its cloud tops make it a consistently conspicuous object, brighter than Mars except when Mars is at its closest opposition distance.

Jupiter rotates in just under ten hours, a shorter period than for any other planet. The value of 9.93 hours quoted in the planetary facts table is the rotation period of the interior, as deduced from the rotation of Jupiter's magnetic field and corresponding periodic variations in radio emissions. The cloud tops that act as Jupiter's visible surface rotate faster than this near the equator, where the cloud top rotation period is 9.84 hours. This corresponds to a westerly (i.e. eastward-blowing) wind of about 150 m per second (540 km per hour)! A notable consequence of Jupiter's rapid rotation is that its shape is perceptibly flattened (Figure 9.1). Its radius measured from centre to pole is only 66,854 km, which is 6.5 per cent less than its equatorial radius.

figure 9.1   Voyager 1 image of Jupiter showing nearly a full disc
the cloud bands run parallel to lines of latitude, and the flattening of the planet's shape towards the poles is apparent
the famous Great Red Spot is visible below the centre of the disc
the small object above right of Jupiter is its innermost large satellite, Io

Jupiter's axial inclination is slight, so that the plane of its equator is very close to the plane of its orbit. Jupiter has 40 known satellites. The eight innermost ones orbit almost exactly in the equatorial plane, but the orbits of the next six are inclined at between 27° and 43°. The rest have orbits inclined at between 147° and 165°, so that their motion is retrograde. This family of satellites and its more important members are described later in this chapter.

Missions to Jupiter

The outer Solar System has been less thoroughly explored than Earth's immediate neighbourhood, but the lure of Jupiter and its amazing satellite system is such that nevertheless it has received

Name	Description	Date of fly-by or operational period
Pioneer 10	fly-by; first close-up pictures, first determination of magnetic field and charged particles, atmospheric measurements	Dec. 1973
Pioneer 11	fly-by; similar to Pioneer 10	Dec. 1974
Voyager 1	fly-by; first detailed pictures of Jupiter's satellites, discovered active volcanoes on Io	Mar. 1979
Voyager 2	fly-by; similar to Voyager 1	July 1979
Ulysses	fly-by as a means to go into polar orbit about the Sun; magnetospheric measurements	Feb. 1992
Galileo	entry probe and orbiter; atmospheric measurements, long duration detailed imaging of satellites	Dec. 1995 (entry probe) – 2003
Cassini	fly-by en route to Saturn	Oct. 2000– March 2001
Jupiter Icy Moons Orbiter	orbiter; imaging, ice-penetrating radar, surface altimetry	after 2014

table 9.1 successful and anticipated missions to Jupiter (all NASA except Ulysses joint with the European Space Agency) (? = exact date unknown)

considerable attention (Table 9.1). The first probes to venture Jupiterwards were Pioneers 10 and 11, which were launched in 1972 and 1973. Their safe crossing of the asteroid belt dispelled fears that this region of space might be so full of meteorite debris that it would be impossible for spacecraft to survive passage through it without unfeasibly robust shielding. Furthermore, by determining the intensity of radiation in the zones of charged particles trapped by Jupiter's magnetic field they enabled the design and trajectories of future missions to be planned within safe limits.

In addition to mapping Jupiter's magnetic field, the Pioneers returned the first close-up images of the planet itself. Probably of more importance than the images were the determinations of pressure, density and temperature in the atmosphere made using infrared detectors and by **radio occultation**, which depended on measuring the way the Pioneers' radio signals faded out as they passed temporarily out of sight behind the planet. In particular these showed that Jupiter's atmosphere is hydrogen-rich but mixed with a significant amount of helium, thereby lending support to the model of Jupiter's growth by scavenging gas from the solar nebula that was described in Chapter 02.

Jupiter and its satellites were explored in more detail by two Voyager probes that flew past in 1979. These recorded detailed time series views of Jupiter's atmospheric circulation and the first ever high resolution views of Jupiter's four largest satellites, revealing active volcanoes on one of them, and also discovering three inner satellites and Jupiter's slender main ring. Each Voyager probe took advantage of Jupiter's gravitational field to alter its trajectory and speed it on its way to Saturn, using the so-called 'gravitational sling-shot' effect otherwise known as a **gravity assist trajectory**. Voyager 2 must rank as the most productive spaceprobe in history, proceeding beyond Saturn to encounters with Uranus and then Neptune (Figure 9.2). As the new millennium dawned both Voyagers were still functioning 76 and 60 AU from the Sun respectively, having overtaken the slower-moving Pioneer probes with which communications ceased in 1997, and returning data on the interplanetary environment. Voyager 1 is expected to be the first human artefact to penetrate the interstellar medium (space beyond the influence of the solar wind) in about 2019.

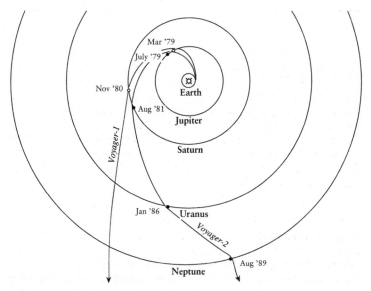

figure 9.2 trajectories of the two Voyager spacecraft
Voyager 1's encounter flung it onwards above (north of) the plane of the Solar System, whereas after Neptune Voyager 2 continued on a course below (south of) this plane

After the Voyagers there was an interval of more than a dozen years before Jupiter received its next visitor from Earth. This was Ulysses, a probe designed to study the Sun. However, in order to get Ulysses into a high-inclination orbit, it was sent first to Jupiter so that by swinging over the planet's north pole the gravitational sling-shot effect would direct it south of the ecliptic plane into an orbit passing over the Sun's south and north poles in turn. This Jupiter swing-past provided a useful opportunity to map the high-latitude parts of Jupiter's magnetosphere, but Ulysses had no imaging system.

For new pictures, the world had to wait for the arrival of Galileo in 1995. This was named after Galileo Galilei, the discoverer of Jupiter's four large satellites, and was the first mission to go into orbit about a giant planet. It swung several times past each of the large satellites in order to produce higher resolution and more complete image coverage than the Voyagers had managed and was still functioning in 2003. Galileo also dispatched a probe into Jupiter's atmosphere. This survived for over an hour reaching nearly 160 km below the cloud tops, where the temperature was 153 °C and the pressure was 22 times that at sea-level on Earth. Entering the atmosphere at 6.5° north of the equator, the probe demonstrated that the westerly wind continues below the cloud tops, with its speed actually increasing to about 190 m per second at the depth where the probe lost contact.

The next mission with Jupiter as its goal will probably be one targetted specifically at Jupiter's icy satellites, listed as Jupiter Icy Moons Orbiter in Table 9.1. This will initially go into orbit around Jupiter itself, possibly in 2014. However, by means of multiple gravity assist fly-bys it will ease itself into a position from which it can be captured into orbit about Callisto, then Ganymede then Europa. There may be an earlier opportunity to collect close-up data on the Jupiter system in March 2007 when a Pluto-bound probe called New Horizons is provisionally scheduled to swing past Jupiter on a gravity assist trajectory.

The interior

Impressive as was the descent of the Galileo entry probe, it penetrated less than a third of a per cent of the way to Jupiter's centre. Our picture of the planet's interior is therefore based on theoretical models of how a large mass of what we understand to be Jupiter's composition should behave, supplemented by insights from the planet's gravity and magnetic fields. As Figure 2.5 shows, the outer layer composed dominantly of molecular hydrogen (mixed with helium) extends below the cloud tops to a depth of about 10,000 km. It would be an oversimplification to refer to the whole layer as 'the atmosphere', because in its lower part (where the pressure exceeds a hundred thousand times the Earth's atmospheric pressure) this hydrogen would behave more like a liquid than a gas. However, the first clear break in properties is believed to occur at about 10,000 km depth, where pressure is about a million times that at sea-level on Earth and temperature is about 6000 °C. Under these conditions the bonds holding together the hydrogen molecules (each consisting of two hydrogen atoms) must break. In this situation, the hydrogen atoms are unbound and the electrons are free to wander through the spaces between the atoms, rather than forming chemical bonds as they would under lower pressure. These are the properties of a molten metal, and hence this layer is described as metallic hydrogen in Figure 2.5. Below this layer Jupiter must have a shell of high pressure ice (probably mostly water in composition), surrounding a core of rock and possibly an inner core of iron. The pressure in the core is probably about 40 million times Earth's atmospheric pressure and the temperature about 17,000 °C. The state of the rocky material in Jupiter's core is unknown. It is probably molten, but the tremendous pressure (20 times that occurring within the Earth) could compress it into a solid. If so, the minerals would be high-density varieties rather than those familiar on Earth. Jupiter's core may look small at the scale of Figure 2.5, but probably contains 10–20 times the mass of the Earth.

Jupiter has by far the strongest magnetic field of all the planets, 20,000 times the strength of the Earth's. It is tilted at 9.6° relative to the planet's rotational axis, and appears to be generated by convection within Jupiter's thick metallic hydrogen layer. This powerful magnetic field holds in place a belt of ionized sulfur centred about the orbit of Jupiter's innermost large satellite, Io, and channels high energy electrons towards Jupiter's poles, where they produce spectacular auroral displays (Figure 9.3).

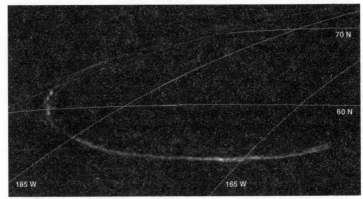

figure 9.3 visible evidence of Jupiter's internally generated magnetic field:
an aurora (glow from ionized atoms) seen by the spaceprobe Galileo on
Jupiter's night-side
lines of latitude and longitude have been superimposed

A notable property of Jupiter, which it shares with Saturn and
Neptune though not Uranus, is that it radiates to space about
twice as much energy as it receives from the Sun. This rate of
heat loss, about seven watts per square metre, is 1000 times the
feasible rate of radiogenic heat production in Jupiter's rocky
core. Instead, it probably indicates that Jupiter is still
contracting in size, converting gravitational potential energy
into heat. If Jupiter had been about thirteen times more massive,
this process would have been sufficient to raise its internal
pressure and temperature to the point where nuclear fusion of
hydrogen begins, and Jupiter would have become more like a
star than a planet.

The atmosphere

It is probably the heat from Jupiter's interior that powers
Jupiter's winds. This is in contrast to the Earth's atmosphere,
where solar heating and condensation of water vapour provide
the impetus for winds. Jupiter's prevailing winds follow a
pattern that is described as 'zonal', meaning confined into
discrete zones of latitude. The fast westerly winds near Jupiter's
equator have already been mentioned. This atmospheric stream
coincides with the pale equatorial zone visible in Figure 9.1. To
both north and south there is a darker belt where the wind
blows in the opposite direction at several tens of metres per

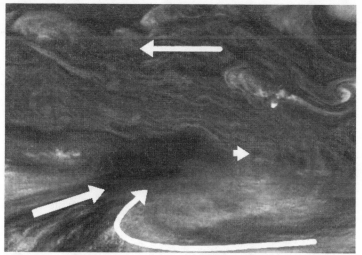

figure 9.4 Galileo image, 34,000 km wide, showing the boundary between westerly (eastward-blowing) winds in Jupiter's pale equatorial zone (lower third of the image) and the easterly winds in the darker belt to its north arrows show wind speed and strength relative to the dark spot, which is a site upon which dry atmosphere is converging and sinking

figure 9.5 Galileo image, 30,000 km across, showing the Great Red Spot in June 1996

the spiral pattern within the Great Red Spot is apparent, reflecting its anticlockwise rotation, and smaller spots and eddies can be seen at the edge of the south tropical zone to its south

this image was recorded in near infrared light, at which wavelength the Great Red Spot is more reflective than its surroundings

second. The winds become super fast again in the direction of the planet's rotation in each of the pale zones at higher latitudes. As might be expected, where zones of opposite flowing winds touch, the shearing between air masses creates complex patterns (Figure 9.4). Generally, the atmosphere is rising in the pale zones and sinking in the dark belts, and effectively Jupiter has a series of several Hadley cells between the equator and the two poles. Jupiter shows very little latitudinal variation in temperature, so these cells must be very efficient at transferring heat from the equator towards the poles. However, the situation is complicated by the fact that so much heat comes from inside the planet too.

figure 9.6 Galileo image, 25,000 km across, showing a remarkable confluence of spiral storm systems on Jupiter in February 1997
the two large white ovals at left and right were storm systems first identified telescopically in 1938
they were rotating anticlockwise, whereas the pear-shaped mass between them was rotating clockwise
these three systems merged into a single feature early in 1998
the spiral storm system in the lower right corner was drifting eastward at 500 km per day relative to its more northerly neighbours

A particularly notable feature of the interaction between competing zonal winds is the generation of eddies where the wind flows in a circular or spiral pattern. The most famous of these is Jupiter's Great Red Spot (Plate 7, Figure 9.5), which has been apparent in telescopic observations since at least 1830 and sits in the dark belt between the pale equatorial zone and the pale south tropical zone. It takes about a week for the Great Red Spot to rotate, and winds around its periphery can reach over 100 m per second. Jupiter has many other storm systems, usually white rather than red, that can individually persist for several years or even decades (Figure 9.6).

In addition to water and the atmospheric constituents listed in the planetary facts table, spectroscopic studies have shown that Jupiter's atmosphere contains traces of gases such as water vapour, hydrogen deuteride (HD, a molecule consisting of an atom of ordinary hydrogen bonded to an atom of 'heavy' hydrogen), ethane (C_2H_6), ethyne (C_2H_2), phosphine (PH_3), carbon monoxide (CO) and germane (GeH_4).

The issue of what imparts the colour to features such as the Great Red Spot is a controversy that has not been resolved despite the Galileo entry probe's success. Compounds of sulfur, phosphorous and carbon have all been suggested. However, we do at least have a reasonable knowledge of the basic composition of the clouds. Clouds occur where conditions of temperature and pressure are such that a constituent in the atmosphere, which may be only a trace (as, for example, water in the atmosphere of Earth and Mars), is more stable as a liquid or solid than as a gas. The whiteness of Jupiter's pale zones is caused by condensation of ammonia-ice (NH_3) as the atmosphere rises and cools towards the cloud top temperature of about −150 °C. In the dark belts, as the atmosphere sinks it becomes hotter, so the ammonia clouds evaporate allowing us to see to deeper and darker levels. About 50 km below the ammonia clouds is a layer of clouds of ammonium hydrosulfide (NH_4HS) at about 0 °C and below this probably a layer of water-ice clouds at about 50 °C. The Galileo entry probe passed through the ammonium hydrosulfide cloud layer, but failed to find the water clouds, probably because it arrived in a dry downwelling zone where the water content of the atmosphere was lower than average.

The Shoemaker Levy 9 impacts

A remarkable opportunity to study Jupiter's atmosphere presented itself with the discovery in March 1993 of comet Shoemaker Levy 9 on a collision course with the planet. The comet had evidently been captured into orbit about Jupiter some decades previously, but this orbit was unstable. It had made an unnoticed close pass by Jupiter in July 1992, only about 21,000 km above the cloud tops, which had tidally disrupted it into a string of about 20 fragments (showing that this comet, like many asteroids, was a loosely bound rubble pile). When discovered these fragments were heading away from Jupiter prior to falling back for a series of 60 km per second impacts over the period 16–22 July 1994.

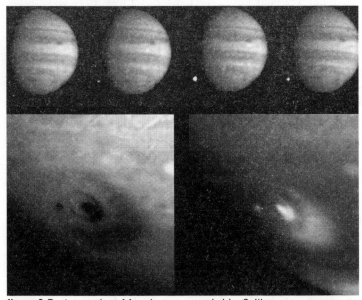

figure 9.7 top: series of four images recorded by Galileo over a seven-second period, showing the beginning, brightening and fading of the fireball produced by the impact of the final fragment of comet Shoemaker Levy 9 on Jupiter on 22 July 1994

bottom: Hubble Space Telescope views recorded on 18 July 1994 showing part of Jupiter's southern hemisphere 105 minutes after a fragment of the comet had collided with the planet

the view on the left was recorded through a green filter, and ejecta that had been thrown out by the impact appears dark

the view on the right was recorded at a near infrared wavelength at which methane absorbs very strongly, so only features high enough to be above most of the atmospheric methane show up brightly

the central spot in each case is the actual impact site, and the crescent shape to its southeast is assymetrically distributed ejecta, that escaped up the angled 'tunnel' drilled through the atmosphere by the comet fragment

the ring about the spot is a shock wave travelling through the atmosphere

The impact coordinates were on Jupiter's night-side just out of direct view from the Earth, but fortunately in a position where Jupiter's rapid rotation would bring the site of each impact into sunlight and direct view from the Earth about 25 minutes later. Even more fortunately, the Galileo spacecraft, still 240 million km and 17 months away from arrival but with a significantly different vantage point, had a direct view of the impact site. Some images of the impacts are shown in Figure 9.7. The Hubble Space Telescope images in this figure show a spot to the left of the most recent impact, which marks the site of an impact that happened 19 hours 43 minutes (almost two rotations) previously. These impact scars were remarkably persistent, some remaining visible for many months before shearing in the atmosphere completely disrupted them. The nature of the dark material, which was probably ejected from some 30 km below the cloud tops remains uncertain, even the professional literature referring to it simply as 'dark stuff'.

Rings

Planetary rings consist of hordes of particles sharing orbits in their planet's equatorial plane, and occur around each of the four giant planets. Their width varies from planet to planet, but in general their thickness is no more than a few tens of kilometres. They generally lie close to their planet, inside what is known as the **Roche limit** where no large solid body could exist because it would be pulled apart by tidal forces. Therefore rings may represent material from that part of an orginal circumplanetary disc that was inside the Roche limit and so could not aggregate into a large body. Alternatively they may owe their origin to a satellite or comet that strayed within the Roche limit and was ripped apart. Some less substantial rings may be supplied by dust sprayed from a nearby satellite into space by impacts or volcanic activity.

The rings of Saturn have been known since the dawn of the telescopic era, although Galileo Galilei himself could not make out what they were. Jupiter's far less spectacular rings were not discovered until Voyager 1's arrival in 1979. Part of the reason for their inconspicuousness is that even the main one is far narrower than Saturn's. Another important factor is that Jupiter's rings reflect less than 5 per cent of the incident sunlight (in other words their albedo is <0.05), presumably because they are made of silicate- or carbon-rich material, whereas Saturn's are icy and reflect up to 80 per cent (albedo up to 0.8).

Jupiter's rings are easier to see in forward-scattered light, i.e. when looking at them towards the Sun (Figure 9.8), than when seen in back-scattered light, which is the only view we ever get of them from Earth. This reveals that they are made mostly of very small particles, about a micrometre in size. There could be plenty of larger chunks too, but the total volume of matter in the rings is probably no more than required to make a single object about a kilometre across.

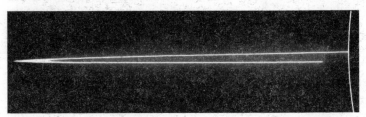

figure 9.8 Jupiter's main ring, as seen by Galileo looking towards the Sun and obliquely down on the ring system
the arc on the extreme right is sunlight scattered by Jupiter's uppermost atmosphere
the nearest part of the ring is invisible because it lies in the planet's shadow
above and below the main ring can be seen a faint halo consisting of particles forced out of the main ring by electromagnetic forces

Micrometre-sized particles can survive in a ring for only about 1000 years before radiation pressure and interactions with the planet's magnetic field disperse them. Therefore either Jupiter's rings are young and ephemeral (which seems improbable) or there is a continual supply of fresh particles.

There are three components to Jupiter's ring system. The main ring is no thicker than 30 km and has an inner radius of 123,000 km and an outer radius of 129,000 km. Its outer edge coincides with the orbit of the tiny satellite Adrastea, whereas the innermost known satellite, Metis, actually orbits within the ring at a radius where the ring shows a marked drop in brightness (not perceptible in the oblique view of Figure 9.8). Apparently unique to Jupiter, there is a 20,000-km-thick halo above and below the main ring that is believed to consist of small particles forced out of the ring plane by electromagnetic forces. Beyond the rim of the main ring lies the so-called gossamer ring that is tenuous in the extreme, but is even thicker than the halo and contains the orbits of the satellites Amalthea and Thebe which are presumably the sources of the particles making up this ring.

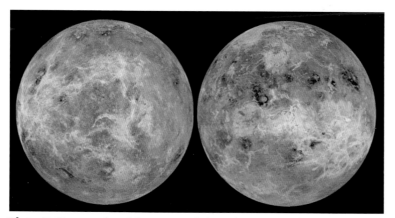

Plate 1 Topography of the western (left) and eastern (right) hemispheres of Venus, as mapped by the Magellan orbiter using radar altimetry. Colour-coding corresponds to height across Venus's 15 km topographic range, with blue lowest and green, brown and white progressively higher.

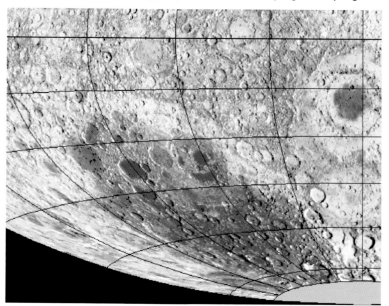

Plate 2 Lunar topography of part of the far side southern hemisphere, colour-coded from Clementine laser altimetry. Colours range from deep purple (8 km below the mean height) through yellow to pink (6 km above mean height), and reveal the extent of the 2500 km diameter South Pole-Aitken basin, which is the largest impact basin on the Moon.

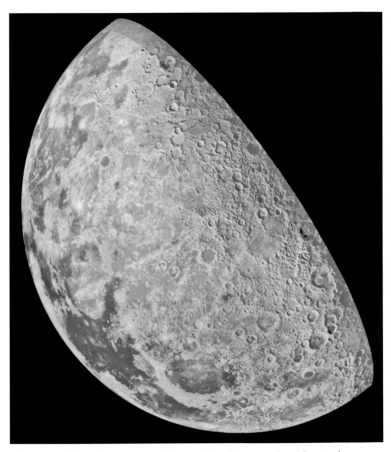

Plate 3 False colour mosaic of part of the lunar northern hemisphere, constructed from images recorded in multiple wavelengths during a lunar fly-by by Galileo. Colour variation from dark blue to orange corresponds to basalt lava flows with differing titanium contents. The lunar highlands appear red. Ejecta from the youngest craters appears pale blue. Apollo 11 landed near the edge of the disc in this view, in the blue area known as Mare Tranquilitatis towards the lower left.

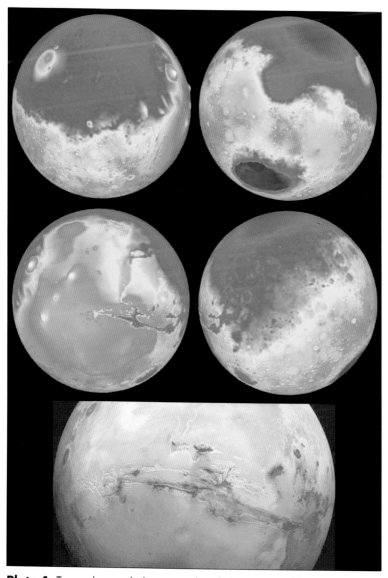

Plate 4 Top: colour-coded topography of Mars, a set of four views showing successive 90 degree rotations. Purple is lowest and white is highest, covering Mars's 30 km height range. Bottom: natural colour image mosaic, constructed as if from an orbital vantage point above the Valles Marineris canyon system.

Plate 5 The surface of Mars at the Mars Pathfinder landing site. The Sojourner rover can be seen analysing a 1 m diameter rock nearby. The peaks on the horizon at top left are less than 2 km away. Parts of the lander and the now-deflated airbags used to cushion its landing are visible in the foreground.

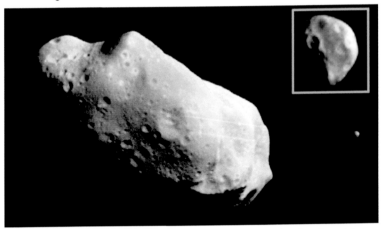

Plate 6 Galileo spaceprobe view of asteroid Ida (56 km long) and its tiny satellite Dactyl. Ida's shape could be a result of collisions. The surface is generally red, a consequence of some kind of space weathering process. In places, young craters expose fresh bluish material whose spectral characteristics suggest Ida is made of the same stuff as common stony meteorites. The inset shows an enlarged view of the most detailed image of Dactyl. Its most prominent crater is 300 m across.

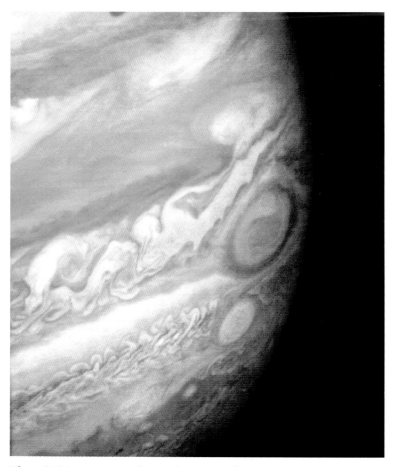

Plate 7 Voyager view of part of Jupiter and its Great Red Spot. This spiral storm pattern is about twice the width of the Earth and takes about a week to rotate.

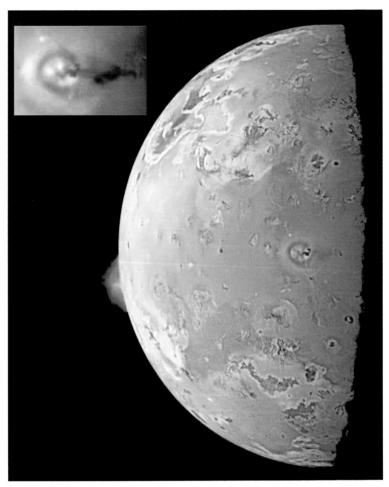

Plate 8 Galileo image of Io recorded on 28 June 1997. There are two eruption plumes visible. One is 140 km high and is seen in profile above the limb; this emanates from a volcano named Pillan Patera. The other plume, from a volcano named Prometheus, is seen from directly above, and lies near the centre of the disc. It is shown enlarged in the inset at upper left; the bluish dark ring is the outine of the plume and this casts a reddish shadow over the surface to its right.

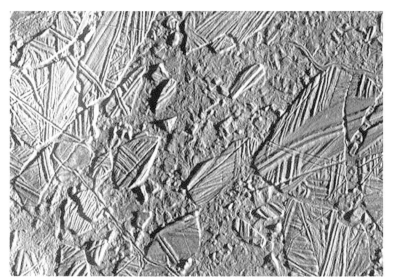

Plate 9 Galileo image of a 60 km wide region of Europa known as Conamara Chaos. The surface of the former bright plains has been broken into rafts or ice floes that drifted apart before the intervening slush or water re-froze. Bright patches on the left are splashes of recent ejecta from a 26 km diameter impact crater 1000 km to the south.

Plate 10 Saturn as seen in January 1998 by the Hubble Space Telescope in reflected near-infrared light, to emphasize atmospheric zoning. The orange clouds near the equator are highest, blue zones near the poles indicate a clear atmosphere down to a lower cloud layer, and green to yellow colours indicate progressively thicker haze hiding this deep layer. Two satellites are visible as tiny specks: Dione on the lower left and Tethys in front of the upper right edge of the disc.

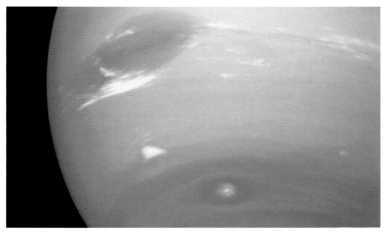

Plate 11 Neptune seen in August 1989 by Voyager 2. At the upper left is the 10,000 km long Great Dark Spot accompanied by rapidly changing white cloud of methane ice crystals, and further south is the Small Dark Spot, which has a bright core.

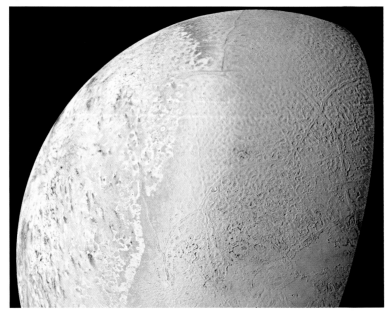

Plate 12 Part of a Voyager 2 image of the sunlit part of the Neptune-facing hemisphere Triton. The south polar cap of bright nitrogen ice is marked by streaks of sooty material erupted from geysers on the left.

Satellites

The satellite system of each giant planet begins closest to the planet with small moonlets associated with the ring system in near-circular orbits, then come larger 'regular' satellites also in near-circular orbits close to the planet's equatorial plane, followed by an outer family of small 'irregular' satellites in elongated, inclined and in most cases retrograde orbits. Most satellites are in synchronous rotation, always keeping the same face towards their planet. This means that one side of a satellite, known as the 'leading hemisphere', faces permanently towards the direction of orbital travel while the 'trailing hemisphere' faces away from the direction of travel. We know comparatively little about most of the inner moonlets and small outer satellites of the giant planets in general, other than that most have densities probably less than that of ice and so are likely to be heaps of rubble and that the inner moonlets are probably the sources of the ring material.

Chapter 02 described how most large satellites of giant planets grew from a disc of gas and dust around each planet. Most inner moonlets are probably fragments of formerly larger satellites that were destroyed by collisions or tidal forces, whereas the irregular outer satellites, especially those in retrograde orbits, are likely to be captured comets or asteroids. These are very dark, and are probably covered by tholins or other carbonaceous material, whatever their internal compositions.

Some basic information about Jupiter's family of satellites is given in Table 9.2. Images of its inner moonlets (Figure 9.9) show that these have the irregular shapes expected of bodies too small for their own gravity to pull them into spheres, the size limit for which appears to be a radius of about 200 km for an icy body. Amalthea and Thebe both have craters that are quite large relative to their own sizes, especially on their leading hemispheres, which is the side most vulnerable to impact damage. Amalthea is dark red in colour, which could be a sulfurous coating derived from its volcanically active neighbour Io, or indicate a tholin-rich surface similar to Jupiter's Trojan asteroids or Centaurs.

Jupiter's irregular satellites are very poorly known. Many of them are recent telescopic discoveries, as can be deduced from the provisional designations that some bear instead of names in Table 9.2. There may be hundreds of others more than 1 km across that still await discovery. The satellite S/1975 J1 was first seen briefly in 1975 but was lost again until 2000, when it was

Name	Distance from planet's centre (km)	Radius (km)	Orbital period (days)	Mass	Density (g/cm³)
Metis	127,960	30 x 17	0.295	1×10^{18} kg	3.0
Adrastrea	128,980	13 x 10 x 8	0.298	0.2×10^{18} kg	3.0
Amalthea	181,300	131 x 73 x 67	0.498	7.5×10^{18} kg	3.1
Thebe	221,900	55 x 45	0.675	0.8×10^{18} kg	3.0
Io	421,600	1821	1.769	8.93×10^{22} kg	3.53
Europa	670,900	1565	3.552	4.80×10^{22} kg	3.01
Ganymede	1,070,000	2,631	7.155	14.8×10^{22} kg	1.94
Callisto	1,883,000	2,403	16.69	10.8×10^{22} kg	1.83
Themisto	7,507,000	4	130.0	?	?
Leda	11,165,000	10	240.9	?	?
Himalia	11,461,000	85	250.6	?	?
Lysithea	11,717,000	18	259.2	?	?
Elara	11,741,000	43	259.6	?	?
S/2000 J11	12,555,000	2.0	287	?	?
Euporie	19,394,000	1.0	553 R	?	?
Euanthe	21,027,000	1.5	620 R	?	?
Harpalyke	21,105,000	2.2	623 R	?	?
Orthosie	21,116,000	1.0	623 R	?	?
Praxidike	21,147,000	3.4	625 R	?	?
Hermippe	21,252,000	2.0	632 R	?	?
Iocaste	21,269,000	2.6	631 R	?	?
Ananke	21,276,000	10	630 R	?	?
Thyone	21,312,000	2.0	632 R	?	?
S/2002 J1	22,931,000	1.5	714 R	?	?
Pasithee	23,029,000	1.0	716 R	?	?
Kale	23,124,000	1.0	721 R	?	?
Chaldene	23,179,000	1.9	724 R	?	?
Isonoe	23,217,000	1.9	726 R	?	?
Eurydome	23,219,000	1.5	721 R	?	?
Erinome	23,279,000	1.6	728 R	?	?
Taygete	23,360,000	2.5	732 R	?	?
Carme	23,404,000	15	734 R	?	?
Aitne	23,547,000	2.0	741 R	?	?
Kalyke	23,583,000	2.6	743 R	?	?
Pasiphae	23,624,000	18	744 R	?	?
Megaclite	23,806,000	2.7	753 R	?	?
Sponde	23,808,000	1.0	749 R	?	?
Sinope	23,939,000	14	759 R	?	?
Callirrhoe	24,102,000	4	759 R	?	?
Autonoe	24,122,000	2.0	765 R	?	?

table 9.2 satellites of Jupiter, not including 21 discoveries made in 2003 orbital distances and periods of the unnamed (provisionally designated) satellites are uncertain (R = retrograde, ? = value unknown)

designated as S/2000 J1 until it became clear that the two objects were the same. It received the name Themisto late in 2002, when ten other discoveries from 1999 and 2000 were also given names. Leda, Himalia, Lysithea, Elara and S/2000 J11 are the only other irregular satellites to have prograde orbits, and those five could be remnants of a single body that broke apart during capture by Jupiter. The outermost satellites all have retrograde orbits. These tend to fall in groups, in which several adjacent orbits have almost the same inclination. Each group is probably the remains of a body that broke apart during capture by Jupiter.

Jupiter has four large satellites, Io, Europa, Ganymede and Callisto, orbiting between its inner and outer satellite families. These were discovered in 1610 by Galileo Galilei, and are known collectively as the galilean satellites. Individually they are substantial bodies and are really worlds in their own rights. Some of their important properties are listed in Table 9.3. The decrease in density outwards from Io through Europa to Ganymede and Callisto reflects the fact that Io is ice-free, Europa has a 100 km thick covering of ice, and Ganymede and Callisto each contain about 40 per cent of ice. This would be a natural consequence of a decrease in temperature with distance away from the hot, young Jupiter within the disc of gas and dust from which these bodies grew. They are described in turn in the following sections.

09

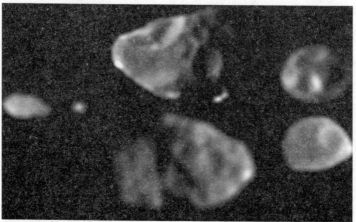

figure 9.9 Jupiter's four innermost known satellites, imaged by Galileo and shown at their correct relative sizes
from left to right: Metis (60 km long), Adrastea (20 km long), Amalthea (247 km long) and Thebe (116 km long)
two views are shown for Amalthea and Thebe: top the leading hemisphere, bottom the trailing hemisphere

	Io	Europa	Ganymede	Callisto
Equatorial radius (km)	1821	1565	2631	2403
Mass (relative to Earth)	0.0149	0.00803	0.0248	0.0181
Density (g/cm³)	3.53	3.01	1.94	1.83
Surface gravity (relative to Earth)	0.18	0.13	0.14	0.13
Composition of surface	rocky	salty ice	dirty ice	dirty ice
Mean surface temperature	−150 °C	−150 °C	−160 °C	−160 °C

table 9.3 planetary facts for the galilean satellites

Io

Of the welter of revelations provided by the Voyager tours of the outer Solar System, the discovery of active volcanoes on Io probably ranks top of the list. Prior to this, most people had assumed that bodies of Io's size, whether rocky like Io or icy like its companions, would be geologically dead like our own, similarly sized, Moon. This is because their small size makes them incapable of having retained enough primordial heat or generating adequate radiogenic heat to keep their lithospheres thin and to drive mantle convection sufficiently close to the surface for melts to escape.

However, it is now realized that the orbital resonance that exists between the three innermost galilean satellites results in tidal heating. For every one orbit completed by Ganymede, Europa completes two and Io four. This means that the satellites repeatedly pass each other at the same points in their orbits, and the consequent internal stresses experienced by the satellites provide a source of heat that keeps their interiors warmer than they would otherwise be. The effect is greatest for Io, which is closest to Jupiter and hence experiences the strongest tidal forces.

There are often more than a dozen volcanoes erupting on Io at any one time. These are identified either by seeing an 'eruption plume' powered by the explosive escape of sulfur dioxide and rising 100–400 km above the surface (Plate 8), or by infrared detection of a hot spot. The record for the highest local temperature is at least 1400 °C.

Io's density is slightly greater than that of the Moon, and it is clear that Io is a dominantly silicate body, like the terrestrial planets. Gravity and magnetic observations by Galileo confirm that it has a dense, presumably iron-rich core below its rocky mantle. Spectral data show that Io's surface is covered by sulfur, sulfur dioxide frost and other sulfur compounds. However, these are no more than thin, volatile, veneers resulting from volcanic activity and the crust as a whole is some kind of silicate rock.

Io has an atmosphere of sulfur dioxide, and atomic oxygen, sodium and potassium. The surface pressure is less than a millionth of the Earth's but nearly a billion times greater than the atmospheric pressure of the Moon or Mercury. Io's atmosphere continually leaks away into space, contributing to a 'cloud' of sodium and potassium that falls inwards towards Jupiter and to a magnetically confined belt of ionized sulfur that stretches right round Jupiter, concentrated around Io's orbit. The atmosphere is replenished by a combination of volcanic activity and collisions onto Io's surface by high-speed ions channelled by Jupiter's magnetic field. When Io passes into the shadow of Jupiter its atmosphere can be seen faintly glowing in an auroral display caused by these same magnetospheric ions impinging on the atmosphere (Figure 9.10).

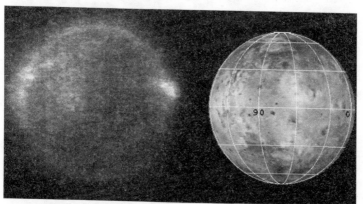

figure 9.10 left: the visible glow over Io's leading hemisphere produced by its atmosphere interacting with high-speed ions channelled by Jupiter's magnetic field

the more brightly glowing patches coincide with volcanic eruption plumes, where the atmosphere is densest

this image was recorded by Galileo when Io was in Jupiter's shadow in May 1998

right: sunlit view of the same hemisphere, with latitude and longitude superimposed (0 degrees longitude is at the centre of the side that always faces Jupiter, and 90 degrees longitude is the centre of the leading hemisphere)

Io's surface is totally dominated by the results of volcanic activity (Figure 9.11). There are lava flows up to several hundred km in length and vast swathes of mostly flat terrain covered by fallout from eruption plumes. Most of the lava flows are now believed to have formed from molten silicate rock, which is often discoloured by a sulfurous surface coating, but there are probably some flows that formed from molten sulfur too. Here and there volcanoes rise above the general level of the plains, and their summits are occupied by volcanic craters (described as 'calderas') up to 200 km across formed by subsidence of the roof of the volcano after magma has been erupted from within. No impact craters are visible, because the volcanic eruptions deposit fresh materials across the globe at an average rate of something like a centimetre thickness per year.

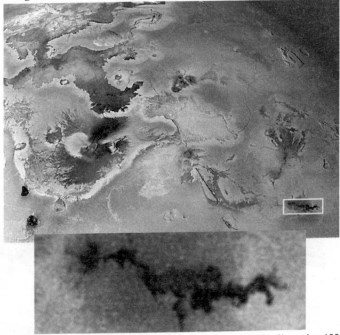

figure 9.11   1500 km wide Galileo image of part of Io recorded in November 1996

an enormous dark lava flow, presumably silicate in composition, occupies the upper left

the two dark spots in the lower left corner and the one to the right of the lava flow are volcanic vents that appear hot in the infrared

the smaller lava flow within the box at the lower right is shown below in a high-resolution view recorded in July 1999

this flow, from a volcanic vent named Zamama, was not present in Voyager images, and therefore formed some time between 1979 and 1996

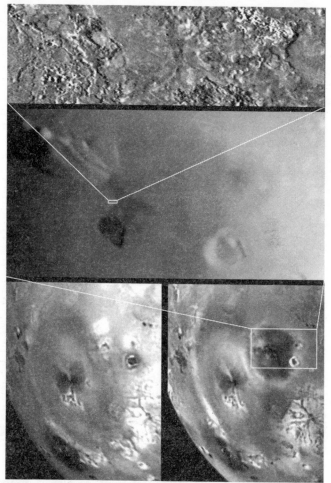

figure 9.12 below: Galileo view of a 1200-km-wide region of Io on
4 April 1997 (left) and 19 September 1997 (right)

between these two dates a short-lived high-temperature eruption plume
occurred at Pillan Patera (see Plate 8) that distributed a 400-km-wide patch
of dark material around the vent, which is the most obvious difference
between the two images

the fainter 1000-km-diameter ring to the left marks the edge of the plume
from a long-lived eruption of Pele volcano

above: successive close-up views of the area indicated, culminating with a
7.2-km-long strip of lava-covered terrain imaged by Galileo from a range of
617 km during a close fly-by of Io on 10 October 1999

Over 500 volcanoes have been identified on Io, and about 100 have been seen to erupt on images from Voyager or Galileo (or both), or by means of infrared telescope studies. The long duration of the Galileo mission enabled many changes on Io's surface to be documented (Figure 9.12). Io's volcanoes appear to be randomly distributed, and Io certainly lacks the kind of well-defined global pattern displayed by the Earth (Figure 5.5). Thus unlike Earth, which gets rid of heat from its interior by plate tectonics, and Venus, where heat escapes by conduction probably punctuated by orgies of resurfacing every half billion years or so, Io's heat escapes by means of hordes of volcanoes. One factor that probably influences the difference between the Earth and Io is that, to maintain a steady state, Io has to lose heat at a rate of about 2.5 watts per square metre, compared to only 0.08 watts per square metre in the case of the Earth. Possibly, the tidal heating experienced by Io is sufficient to keep a large fraction of its mantle partially molten.

Europa

For all Io's majesty, its neighbour Europa excites the greatest scientific interest. Europa is a transitional world, with a density almost in the terrestrial planet league but an exterior that is icy down to a depth of about 100 km. It is not known whether the ice is solid throughout, or whether its lower part is liquid, which raises the fascinating possibility of a global ocean sandwiched between the solid ice and the underlying rock. Gravity data from Galileo show that, like Io, Europa has a dense, presumably iron-rich core (about 620 km in radius) below its rocky mantle. Europa has its own magnetic field, but it is not clear whether this is generated by convection within a liquid core or within a salty ocean beneath the ice.

Europa has a highly reflective surface with an albedo of about 0.7, and it has been known since the 1950s from spectroscopic studies that its composition is essentially that of clean water-ice. More detailed recent observations by Galileo and the Hubble Space Telescope reveal some regions where the ice appears to be salty, and also the presence of molecular oxygen (O_2) and ozone (O_3). The oxygen and ozone are thought to result from breakdown of water molecules in the ice because of exposure to solar ultraviolet radiation and charged particles. The hydrogen so liberated would escape rapidly to space, which has been observed on Ganymede though not on Europa. It is not known whether the oxygen and ozone detected on Europa constitute an extremely tenuous atmosphere or are mainly trapped within the ice.

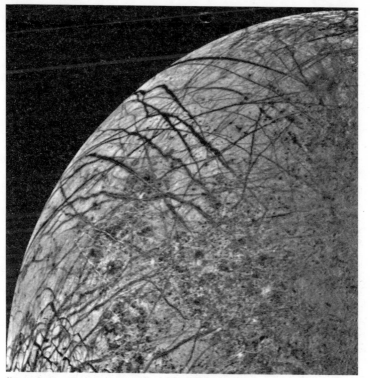

figure 9.13 Voyager 2 view of part of Europa
bright plains cover most of this view, with mottled terrain occupying the
lower right

Europa's surface is relatively smooth and much younger than
that of other icy satellites, to judge from the paucity of impact
craters. This demonstrates that Europa experiences a significant
amount of tidal heating, though less than Io. Images at Voyager
resolution, such as Figure 9.13, show bright plains criss-crossed
by a complex pattern of cracks filled by darker ice. There are
several places where the pattern of these bright plains becomes
blotchy, and these were dubbed 'mottled terrain' by the Voyager
investigators.

The high resolution images sent back by Galileo show that the
bright plains are amazingly complex in detail (Figure 9.14),
being composed of a pattern of straight or slightly curved ridges,
each usually bearing a central groove. The appearance of these
parts of Europa has been described as resembling the surface of
a ball of string; an apt description but not much help in trying
to decipher how the surface was created. Each grooved ridge

could represent a fissure that was the site of an eruptive episode, when some kind of icy lava was erupted.

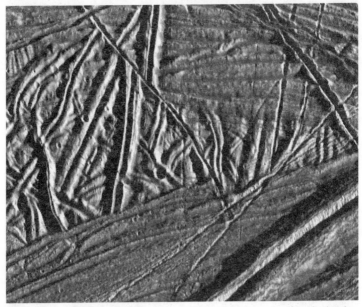

figure 9.14 15-km-wide Galileo view of part of Europa, showing the level of complexity revealed in Europa's bright plains by Galileo's high resolution the relatively smooth area in the lower third of this view corresponds to a dark crack at Voyager resolution, whereas the complexly rigid surface elsewhere looks bright and featureless on images like Figure 9.13 the large ridge cutting across the lower right is 300 m high there is a 300-m-diameter impact crater near the centre and several smaller ones, so although young by Solar System standards this surface is likely to be at least several million years old

It may seem strange to read of 'icy lava', but Solar System ice shares many important properties with the silicate rock that melts to produce lava on the terrestrial planets. Unless the ice is absolutely pure water, these properties include:

- existing in the solid state as intergrown crystals of differing compositions
- rock-like strength and rigidity under prevailing surface conditions contrasted with the ability to flow slowly and act as an asthenosphere at depth
- the capacity to partially melt, yielding melts different in composition to the starting material.

Planetary scientists often use the term **cryovolcanism** to denote icy rather than silicate volcanism. Although by far the most abundant component in Europa's ice is water, it is likely to be contaminated by various salts (such as sulfates, carbonates and chlorides of magnesium, sodium and potassium) resulting from chemical reactions between water and the underlying rock. The spectroscopic data for Europa are most consistent with the salt-rich areas of surface being rich in hydrated sulfates of magnesium or carbonates of sodium, but could also indicate the presence of frozen sulfuric acid. Contaminants such as these could make any melt liberated from the ice behave in a much more viscous (i.e. less runny) manner than pure water. If erupted as a liquid this type of lava would not necessarily spread very far before congealing, especially if confined by a chilled skin of the sort likely to form upon exposure to the vacuum of space in the cold outer Solar System. Contaminants also allow the ice to begin to melt at a much lower temperature than pure water-ice: salts can depress the melting temperature by a few degrees and sulfuric acid by as much as 55 degrees.

Maybe, then, Europa's ridges are simply highly viscous cryovolcanic flows fed from their central fissures. Alternatively, the cryovolcanic lava may not have flowed across the ground at all: it could have been flung up from the fissure in semi-molten clods by mild explosive activity, like a 'fire fountain' from a volcanic rift on Earth, and fallen back to coalesce as a rampart on either side of the fissure.

Irrespective of refinements such as this, it seems inescapable that each fissuring event must represent the opening of an extensional fracture in the crust. This cannot happen across an entire planetary body unless the globe is expanding, which seems highly unlikely. Therefore there must be some regions on Europa where surface has been destroyed at a rate sufficent to match the crustal extension elsewhere. Likely candidates for this on Europa are regions described as 'chaos', and part of one of Europa's chaos regions is shown in Plate 9. Here typical-looking bright plains crust has been broken into rafts that have drifted apart, maybe because an underlying liquid ocean broke through to the surface. The areas intervening between rafts are a jumbled mess reminiscent of re-frozen sea-ice on Earth. Some rafts can be fitted back together, but it is apparent that many pieces of the 'jig-saw puzzle' are missing. Perhaps these missing pieces have sunk or been dragged down beneath the surface.

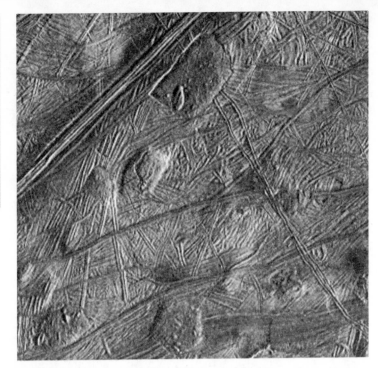

figure 9.15 60-km-wide Galileo view of part of Europa
this region, mapped as mottled terrain on Voyager images, is revealed to be
conventional bright plains with 'ball of string' texture disrupted by a number
of domes, some of which have pierced the former surface

Regions like Plate 9 appeared as mottled terrain on Voyager images, but so did the region shown in Figure 9.15. Here, the surface of what was formerly normal looking bright terrain has been forced up into a number of domes up to 15 km across. Presumably this is because of the rise of pods of molten or semi-fluid low density material (described geologically as 'diapirs') toward the surface. In some cases the upwelling pod has actually ruptured the surface, to form a small chaos region bearing raftlets of surviving crust.

Although we understand all too little about the processes that have shaped Europa's surface, it is clear that it has had a complicated history. We cannot tell for sure how old each region of surface is, but there are abundant signs that there is, or has been, a liquid zone below the surface ice. A salty ocean below several km of ice is not necessarily a hostile environment for life, and indeed life down there could be much richer and complex than anything that is likely to have survived on Mars. In the depths of the Earth's oceans there are whole living communities that are independent of photosynthetic plants (requiring sunlight to live) and depend instead on bacteria-like microbes that make a living from the chemical energy supplied by springs of hot water ('hydrothermal vents') on the ocean floor.

Given that Europa is tidally heated, we can imagine zones where water is drawn down into the rocky mantle, becomes heated, dissolves chemicals out of the rock, and emerges at hydrothermal vents surrounded by life. This may sound far-fetched, but hydrothermal vents are now mooted as the most likely venue for life to have begun on Earth, so if it could happen here then why not on Europa too? It is the possibility of a life-bearing ocean below the ice that is the main driver for plans for the future exploration of Europa.

The proposed Europa Orbiter mission (Table 9.1) has primary goals of verifying the existence of an ocean below the ice, and identifying places where the ice is thin enough for future landing missions to release robotic submarines ('hydrobots') under the ice so that they can go hunting for hot springs and their attendant life. It will achieve this by gravity studies, by using an altimeter to determine the height of the tide raised on Europa by Jupiter (only 1 m if the ice is solid throughout, but about 30 m for 10 km of ice overlying a global ocean), and by using ice-penetrating radar to map ice thickness. High resolution conventional images will identify sites of recent eruptions.

Insights into how best to go about the future exploration of Europa's ocean will probably be gained by study of Lake Vostok, a lake of 10,000 square km beneath 4 km of ice that was discovered in Antarctica in 1996. Its waters may have been isolated from the surface for as long as 30 million years, and may host a unique ecosystem. The development of technology to drill through the ice and explore Lake Vostok, and also the adoption of protocols to avoid biological contamination of this special environment should both provide valuable experience.

Ganymede

Ganymede is the largest planetary satellite in the Solar System, being bigger (though less massive) than the planet Mercury. It is shown in comparison with its outer neighbour Callisto in Figure 9.16. Although these two satellites are similar in size, with bulk

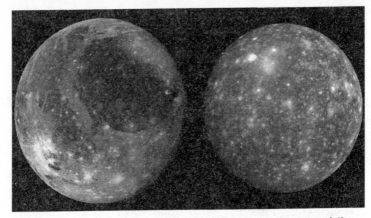

figure 9.16 Ganymede (left) and Callisto (right) seen at the same relative scales
the distinct pale and dark terrains on Ganymede are obvious
Callisto has no such distinction, and is dark all over (its brightness has been exaggerated here to help it show up)
the bright spots on both bodies are relatively young impact craters, made prominent by their surrounding blankets of pale ejecta

densities implying a roughly 60:40 rock:ice mixture in each, Galileo indicated that Ganymede is a fully differentiated body, with an iron inner core (filling 22 per cent of its radius), a silicate outer core (filling 55 per cent of its radius) and an icy mantle, but that Callisto is only weakly differentiated. The difference in evolution between these two bodies is probably because Ganymede was formerly subject to much more intense tidal heating than it receives today, whereas Callisto has never experienced much heating, tidal or otherwise. Ganymede has a magnetic field with about 1 per cent the strength of the Earth, which could be generated in the core or in a salty ocean deep within the ice layer.

With an albedo of about 0.45, Ganymede's surface is darker than Europa's. Spectroscopic studies show that it is dominantly water-ice, with scattered patches of carbon dioxide ice, and that the darkening is caused by silicate minerals (probably in the form of clay particles) and tholins. The darkening is at least partly attributable to the much greater age of Ganymede's surface, allowing more time for the action of solar radiation to produce tholins and for silicate grains to become concentrated in the regolith by the preferential loss of ice during impacts.

There are also faint traces of oxygen and ozone, apparently trapped within the ice as suggested for Europa, and Galileo found an extremely tenuous (and continually leaking) atmosphere of hydrogen that is presumably the counterpart to the oxygen produced by the breakdown of water molecules.

It is obvious even on images of the resolution of Figure 9.16 that there are two distinct terrain types on Ganymede, one darker than the other. At higher resolution (e.g. Figure 9.17) it is apparent that the pale terrain must be younger than the dark terrain, because belts of pale terrain can be seen to cut across pre-existing tracts of dark terrain. However, the density of impact craters on both terrain units shows that each must be very old, perhaps as much as 3 billion years.

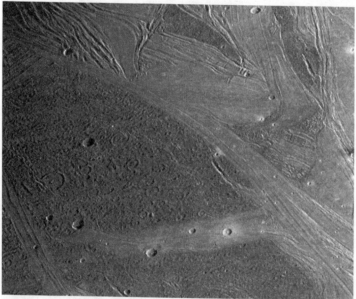

figure 9.17 600-km-wide Galileo view of part of Ganymede, showing belts of pale terrain cutting dark terrain (lower left) and several generations of cross-cutting pale terrain (notably near the top)
there are plenty of young, crisp-looking impact craters on both terrain types, and the dark terrain also has large numbers of more subdued-looking older craters

Considered in more detail, it becomes clear that multiple episodes of pale terrain generation are required to explain the cross-cutting relationships between belts of pale terrain. Images like Figure 9.17 and others at a higher resolution, show the complex grooved nature of the pale terrain and hint that each belt of pale terrain might require just as complex a series of events to explain it as parts of the bright plains on Europa.

This is not to say that the same processes were involved. One important difference between the two bodies is that on Europa there is abundant evidence of the two halves of a split tract of terrain having been moved apart to accommodate the new surface that has formed in between, whereas on Ganymede signs of lateral movement are scarce. Ganymede's pale terrain appears to occupy sites where the older surface has been dropped down by fault movements, allowing cryovolcanic fluids to spill out. It remains a matter of debate whether the grooves within each belt of pale terrain represent constructional cryovolcanic features or reflect subsequent deformation, perhaps related to underlying faults.

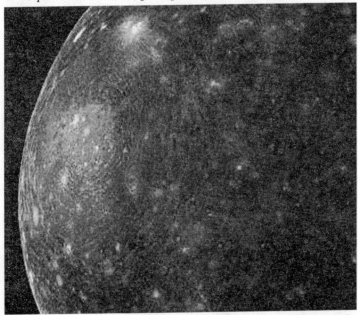

figure 9.18 Galileo view of a large part of Callisto showing its dark, heavily cratered surface

the younger impact craters are the most prominent, because their ejecta blankets have not yet become radiation-darkened

towards the left is a multiringed impact basin named Valhalla

the 600-km-diameter pale zone in Valhalla's centre marks the site of the original crater, but this was too large a structure for the lithosphere to support and it has long since collapsed

this is surrounded by rings of concentric fractures, the outermost having a diameter of 4000 km

Callisto

Callisto, the outermost galilean satellite, has a dark heavily cratered surface (Figure 9.18) with an albedo of only 0.2. It is the only one of its family to resemble what was expected before the role of tidal heating became appreciated. Callisto is not currently in orbital resonance with any of its neighbours, and shows no clear sign of having experienced tidal heating in the past.

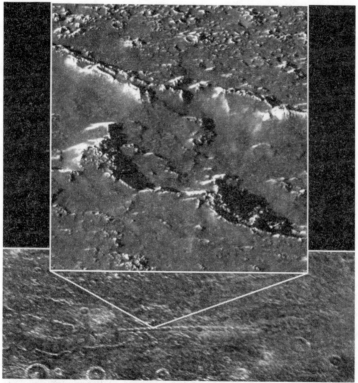

figure 9.19 a chain of 25 10-km-diameter craters on Callisto believed to result from the serial impacts of fragments of a tidally disrupted comet
bottom: Voyager image showing the whole chain in context
top: Galileo view, showing an oblique close-up view of parts of three overlapping craters in this chain

Despite this, Callisto is far from boring. Its weakly differentiated structure has already been remarked upon, and appears to consist of an ice-rock mixture throughout except for an ice-rich crust. There could possibly be a rocky core occupying up to 25 per cent of Callisto's radius, but an iron core would seem to be ruled out by the Galileo gravity data. Therefore it is surprising that Galileo found that a magnetic field is induced within Callisto by its passage through Jupiter's magnetosphere. Given the lack of an iron core, the only reasonable way to explain this magnetic field is to appeal to a salty (and therefore electrically conducting) ocean at least 10 km thick and no more than 100 km below the surface. A liquid layer at such a shallow depth seems incompatible with the undeformed, ancient and heavily cratered nature of Callisto's surface, so here is yet another mystery awaiting resolution.

However, some features on Callisto that used to be a mystery are now understood thanks to comet Shoemaker Levy 9. These are linear chains of overlapping craters (Figure 9.19). It is now agreed that each chain is the scar of impacts by fragments of a different comet, each of which hit Callisto on its outward path immediately after having been tidally disrupted during a close passage by Jupiter. There are about a dozen of these, mostly on Callisto's leading hemisphere which is the side most exposed to the risk of such impacts.

High resolution views like the Galileo image in Figure 9.19 reveal a surprising paucity of craters less than 1 km across. Given the number of larger craters that are present, it is inconceivable that smaller craters have not formed in even greater abundance. Therefore something must be acting to remove them. The Figure 9.19 Galileo image contains clues as to what might be going on. Hilltops and some slopes are markedly brighter than the dark surface from which they crop out. This can be explained if the bright surfaces are relatively clean ice whereas the dark surfaces are covered by a regolith that is enriched in rock debris.

This may be a result of the continual bombardment of the surface by charged particles and micrometeorites. We have seen that this contributes to the breakdown of water molecules, but it can also simply vaporize the ice. On Callisto, those water vapour molecules that do not escape to space or become split into hydrogen and oxygen will recondense on any icy surface that they bump into. This condensation process is at its most efficient on hilltops because these are exposed to the sky all round and so are slightly colder than the surrounding area. Once the brightness difference is established, the temperature contrast gets reinforced because the brighter surfaces will absorb less solar warmth than the darker ones. Therefore, over time frost migrates towards the bright hilltops and other exposed areas, and the residual silicate dust becomes progressively concentrated in the places from where the ice is being selectively removed. Perhaps this process is capable of eroding the rims of small craters faster than the average rate at which such craters are forming.

Another erosional process that occurs on Callisto is landslips, as can be seen in Figure 9.20, where the inner part of a crater rim has collapsed.

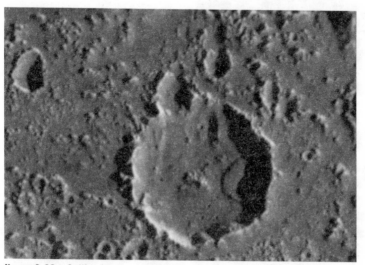

figure 9.20 Galileo image showing a 12-km-wide crater on Callisto with a landslip extending a third of the way across its floor from its eastern rim note the degraded morphologies of many of the smaller craters, and the relative brightness of the hilltops compared to the dustier low ground

10

Saturn

In this chapter you will learn:

- about Saturn, a smaller version of Jupiter, but with a much more spectacular ring system. It has a diverse family of satellites, including one with a dense atmosphere.

Planetary facts	
Equatorial radius (km)	60,268
Mass (relative to Earth)	95.16
Density (g/cm³)	0.69
Surface gravity (relative to Earth)	0.92
Rotation period	10.66 hours
Axial inclination	26.7°
Distance from Sun (AU)	9.55
Orbital period	29.42 years
Orbital eccentricity	0.056
Composition of surface	gassy
Mean cloud-top temperature	−180 °C
Composition of atmosphere	hydrogen (97%), helium (3%), methane (0.5%), ammonia (0.01%)
Number of satellites	33

Saturn is unquestionably the most beautiful sight in the Solar System (Plate 10), and even a small modern telescope will reveal its ring system. With his primitive telescope in 1610 Galileo Galilei could make out that there was something strange about Saturn, but it was Christiaan Huygens in 1659 who first realized that the planet is encircled by disc-like rings.

Rotation and orbit

Earth overtakes Saturn every 378 days, so that opposition occurs about two weeks later each year. Saturn's axial inclination of nearly 27° is such that our view of the rings (which are in its equatorial plane) depends very much on where Saturn is in its orbit. At two periods per orbit Saturn presents its rings edge on as seen from the Earth, at which time they disappear from view. When the rings are invisible Saturn appears somewhat fainter in the sky than when the rings are tilted towards us, and this is a more important factor in determining Saturn's brightness in the sky than the Earth–Saturn distance at opposition. Even so Saturn always rivals the very brightest stars.

Saturn rotates almost as rapidly as Jupiter, but because of its lower density its shape is even more flattened, as is apparent in Plate 10. In fact, Saturn's polar radius is 10 per cent less than its equatorial radius. The rotation period of the atmosphere near the equator exceeds the rate of internal rotation even more than in Jupiter's case, corresponding to winds of about 500 metres per second.

Saturn has 33 proven satellites, including two tiny ones discovered by the Cassini spaceprobe in 2004. Like the Jupiter system, the inner satellites have near circular prograde orbits lying close to the equatorial plane, but only one is comparable in size with Jupiter's galilean satellites. The outer satellites have more eccentric, inclined and commonly retrograde orbits.

Missions to Saturn

Name	Description	Date of fly-by or operational period
Pioneer 11	fly-by; first close-up pictures, first determination of magnetic field and charged particles, atmospheric measurements	Sept. 1979
Voyager 1	fly-by; first detailed pictures of Saturn's satellites	Nov. 1980
Voyager 2	fly-by; complementary coverage to Voyager 1	Aug. 1981
Cassini	orbiter and Titan entry probe (Huygens); optical and radar imaging	July 2004– June 2008

table 10.1   successful and anticipated missions to Saturn (all NASA except Cassini joint with the European Space Agency)

The first spaceprobe to reach Saturn was Pioneer 11 in 1979 (Table 10.1), which gave a taste of what was to come shortly afterwards with the arrival of the faster moving Voyagers 1 and 2. Voyager 1 was sent behind Saturn's largest satellite, Titan, so that radio occultation could be used to characterize Titan's atmosphere, and then over Saturn's south pole to give similar information on the planet's atmosphere. The resulting gravity assist flung Voyager 1 northwards out of the ecliptic plane so that it could make no further planetary encounters, but Voyager 2 stayed in the ecliptic plane and followed a gravity assist trajectory round Saturn's equatorial region to speed it on its way to Uranus.

Future on-the-spot studies of Saturn must await the arrival of Cassini. This was launched in October 1997 and arrived in orbit around Saturn in July 2004. Its main target is not Saturn itself, but its satellites. Titan in particular will receive close scrutiny, using, like Magellan at Venus, a radar to construct images of its surface through the clouds. Moreover it will send a probe called Huygens (named in honour of the discoverer of the rings, who also discovered Titan itself in 1655) down to the surface by parachute in January 2005.

The interior

Saturn's internal structure is compared with that of the other giant planets in Figure 2.5. As for Jupiter, Saturn's internal temperature and pressure are thought to be sufficient to produce a zone of metallic hydrogen surrounding the icy core, and Saturn too radiates to space more energy than it receives from the Sun. However, in Saturn's case the likely source of most of this heat is differentiation occuring by segregation of helium towards the bottom of the molecular hydrogen layer. Such a mechanism is thought impossible within Jupiter where this zone would be too vigorously stirred by convection. The pressure and temperature in Saturn's core are less than in Jupiter, but still considerable: about 10 million times Earth's atmospheric pressure and at least 9000 °C.

Saturn has a magnetic field about 600 times the strength of the Earth's. This is like a scaled down version of Jupiter's field and produces similar polar aurorae, except that it is exactly aligned with the planet's axis of rotation. Like Jupiter's field, it is thought to be generated by convection within the metallic hydrogen layer.

The atmosphere

Saturn's atmosphere is similar to Jupiter's, having prevailing winds blowing from west to east parallel to the equator, with Hadley circulation superimposed. The equatorial wind speed is considerably faster than on Jupiter, and exceeds 400 m per second for 10° either side of the equator. The main differences are that dark (sinking) belts and pale (rising) zones are not so apparent, there is less colour, and the major storm systems are shorter-lived.

In addition to the principal atmospheric constituents listed in the planetary facts table, traces of ethane (C_2H_6), ethyne (C_2H_2), and phosphine (PH_3) have been detected spectroscopically. Like Jupiter's, Saturn's visible surface comprises clouds of ammonia-ice, presumed to be underlain in turn by layers of ammonium hydrosulfide (NH_4HS) and then water-ice clouds. However, because of Saturn's gentler gravity, the height gap between each cloud layer is nearly twice what it is in Jupiter's atmosphere.

figure 10.1 Hubble Space Telescope view of a globe-encircling Saturn storm, recorded on 9 November 1990
this feature first developed as a simple white spot 46 days previously

Figure 10.1 shows one of the largest storms seen on Saturn, of a kind that seems to occur about once every 30 years coincident with the northern hemisphere summer. This event began as a discrete white spot near the equator that spread to encircle the planet within a month, and then took several months to fade from view. It was the most spectacular outburst on Saturn since 1933.

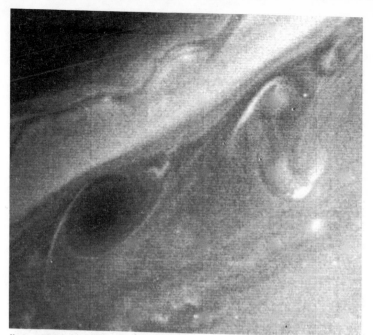

figure 10.2 Voyager 2 image showing a 16,000-km-wide region of Saturn
the dark spot is centred at about 42° north and sits in a zone of 20 m per
second west-blowing winds
however, winds blow in the opposite direction at 150 m per second along the
wavy edge of the belt of white clouds further north

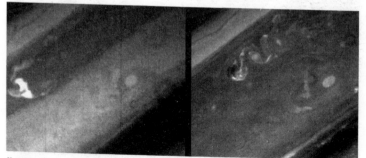

figure 10.3 Voyager 1 (left) and Voyager 2 (right) views of a 30,000 km
wide region to the north of Saturn's equator, recorded on 5 November 1980
and 21 August 1981
the 2500 km white oval in the centre right of each view (fainter in the
Voyager 1 image) is the same storm system
note the difference between these two dates in the turbulent airstream
10,000 km to its north

More normal storms on Saturn appear in Figure 10.2, which shows a 4000-km-diameter dark rotating storm system generated near an interface between opposite airstreams, and Figure 10.3 which shows a 2500 km diameter white oval that persisted for 320 days between the Voyager 1 and Voyager 2 encounters with Saturn.

Rings

Saturn's rings are more substantial than Jupiter's, containing enough material to make a single object about 100 km across. They are made of highly reflective material, which explains their spectacular appearance as seen in Figure 10.4 and Plate 10. The easily visible extent of the rings extends from about 75,000 km to about 137,000 km from the planet's centre, but is less than 100 m thick. There is one clearly visible gap in the rings at a radius of about 120,000 km. This is Cassini's division (named after Giovanni Cassini who discovered it in 1675) and it separates the outer, or A, ring from the inner, or B, ring which is the brightest and most optically dense part of the rings. Within the B ring lies the more transparent C ring, known to telescopic observers as the crepe ring. This can be made out against the blackness of space in

figure 10.4 Voyager 1's farewell view of Saturn, after its August 1981 fly-by the rings' shadow can be seen on the planet, as well as the planet's shadow on the rings, and it is possible to see the planet through all but the brightest part of the rings

the nightside hemisphere of Saturn is faintly illuminated by sunlight reflected from the rings

Figure 10.4 and Plate 10 but is sufficiently transparent to allow Saturn to shine through it.

The relatively slow cooling rate of the A, B and C rings as they pass into Saturn's shadow, the ability of ground-based radar to obtain a signal reflected by the rings, and the effect of the rings on Voyager radio signals all combine to show that they are made of chunks ranging from about 1 cm to about 5 m in size. This is considerably larger than the bulk of Jupiter's ring material. Spectroscopically, Saturn's ring material can be identified as mostly water-ice, but darkened and in most places reddened by radiation damage or by contaminants such as dust.

figure 10.5 Voyager 2 view of part of Saturn's C ring (filling most of the frame) and part of the inner zone of the B ring, showing the concentric gaps and variations in brightness within them
this is a foreshortened, oblique, view covering 20,000 km from the innermost to outermost part of the rings shown

Saturn has other rings too narrow or tenuous to make out in Figure 10.4 and Plate 10. There are two tenuous but broad rings: the D ring, which extends from the inner edge of the C ring halfway down to the cloud tops, and the E ring which is the outermost known, extending from about 180,000 km to about 500,000 km, and expanding to a thickness of about 30,000 km at its outer edge. There are also two narrow rings between the A ring and the E ring. These are the F ring (discovered by Pioneer 11), which is 30–500 km wide and lies only 4000 km beyond the A ring, and the G ring (discovered by Voyager), a tenuous 8000-km-wide feature at a radius of 170,000 km.

This description may already seem complicated enough, but there is more to come. Cassini's division is merely the widest and most prominent of a vast number of gaps within the A, B and C rings (Figure 10.5). Many of these occur at radii where any ring particles would be in orbital resonance with one of Saturn's satellites, for example the inner edge of Cassini's division corresponds to a 2:1 orbital resonance with Mimas. In contrast, Encke's division, a gap near the outer edge of the A ring that can be made out from Earth with a good telescope, is swept clear because Saturn's smallest known inner moonlet satellite, Pan, which was discovered on Voyager images, orbits along the track of the gap. Yet other gaps are controlled by more complex dynamic interactions. The narrow F ring is kept in place by a pair of 'shepherd' satellites, Prometheus that orbits just inside it and Pandora that orbits just outside it. From time to time these distort the F ring, changing its shape from a featureless ribbon to a twisted braid. On top of all this, the concentric symmetry of the B ring is sporadically broken by dusty radial spokes.

Satellites

Saturn has a diverse family of satellites (Table 10.2). Most of these are highly reflective icy bodies. The four innermost ones were discovered on Voyager images. Discoveries of four additional satellites near the F ring were announced as a result of telescopic observations during Earth's 1995 ring plane crossing, when the absence of the usual glare from the rings improved the chances of detecting small faint objects. However, two of these are now dismissed as temporary clumps of material within the F ring, and the other two turned out simply to be Atlas and Prometheus whose orbits had either not been adequately defined by the Voyager observations, or had altered slightly. However 13 genuine tiny new outer satellites were discovered in 2000 and 2003 as a result of telescopic surveys.

Name	Distance from planet's centre (km)	Radius (km)	Orbital period (days)	Mass	Density (g/cm³)
Pan	133,583	10	0.575	?	?
Altas	137,640	18 × 14	0.602	?	?
Prometheus	139,640	74 × 34	0.613	1.4×10^{17} kg	0.27
Pandora	141,700	55 × 31	0.629	1.3×10^{17} kg	0.42
Epimetheus	151,422	69 × 53	0.695	5.5×10^{17} kg	0.63
Janus	151,472	99 × 76	0.695	2.0×10^{18} kg	0.65
Mimas	185,520	199	0.942	3.7×10^{19} kg	1.14
S/2004 S1	194,000	1.5	1.997	?	?
S/2004 S2	211,000	2	1.14	?	?
Enceladus	238,020	249	1.370	6.5×10^{19} kg	1.00
Tethys	294,660	529	1.888	6.1×10^{20} kg	1.00
Telesto	294,660	15 × 8	1.888	?	?
Calypso	294,660	15 × 8	1.888	?	?
Dione	377,400	560	2.737	1.1×10^{21} kg	1.50
Helene	377,400	16	2.737	?	?
Rhea	527,040	764	4.518	2.3×10^{21} kg	1.24
Titan	1,221,850	2575	15.95	1.34×10^{23} kg	1.88
Hyperion	1,481,100	165 × 113	21.28	?	?
Iapetus	3,561,300	720	79.33	1.6×10^{21} kg	1.0
Kiviuq	11,300,000	7	449	?	?
Ijiraq	11,500,000	5	453	?	?
Phoebe	12,952,000	115 × 105	550.5 R	?	?
Paaliaq	15,200,000	10	687	?	?
Albiorix	15,600,000	13	738	?	?
Skadi	15,700,000	3	731 R	?	?
Erriapo	17,500,000	4	858	?	?
Siarnaq	18,160,000	16	893	?	?
Tarvos	18,250,000	7	924	?	?
Mundilfari	18,710,000	3	951 R	?	?
S/2003 S1	18,720,000	3	956 R	?	?
Suttung	19,460,000	3	1016 R	?	?
Thrym	20,380,000	3	1087 R	?	?
Ymir	23,100,000	8	1312 R	?	?

table 10.2   satellites of Saturn (R = retrograde, ? = value unknown)

figure 10.6 Saturn's innermost satellites (excluding Pan which has never been clearly seen) shown at their correct relative sizes
from left to right: Atlas, Pandora (above) and Prometheus (below), Janus (above) and Epimetheus (below)
Janus is 99 km long
the shadow of Saturn's F ring can be seen falling across Epimetheus

Five of the small inner moonlets are shown in Figure 10.6. Atlas orbits just outside the A ring. Prometheus and Pandora are the F ring shepherds referred to above whose combined gravitational attraction acts to confine the ring and sometimes to twist it into a braid. Janus and Epimetheus share virtually the same orbit as each other, in 7:6 orbital resonance with the outer edge of the A ring. As is only to be expected, like Jupiter's inner satellites they are irregular in shape and scarred by many craters.

Some of Saturn's other small satellites are shown in Figure 10.7. These are a particularly diverse assortment. Calypso and Telesto share the same orbit as the much larger satellite Tethys. Telesto travels 60° ahead and Calypso 60° behind the larger body, bearing the same relationship to it as do the Trojan asteroids to Jupiter (Chapter 08). Similarly Helene, which bears one remarkably large crater for its size, occupies the leading Trojan point of the larger satellite Dione. No counterpart to Helene has been found in Dione's trailing Trojan point, but if there is one this and 'Trojan' companions to other larger satellites may be located by the Cassini mission.

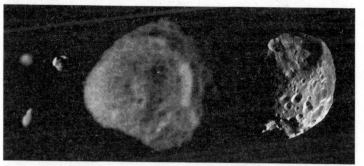

figure 10.7 some other small satellites of Saturn shown at their correct relative sizes

from left to right: Telesto (above) and Calypso (below), Helene, Hyperion and Phoebe
Hyperion is 185 km long

(Voyager images except for Cassini image of Phoebe.)

Hyperion is Saturn's largest irregularly shaped satellite. It is dark and somewhat red, and is a strong candidate for being a fragment from the interior of a larger body that was destroyed by a collision. Hyperion is remarkable in that as it progresses round its orbit it tumbles in a seemingly random way, with both its rotation period and axis of rotation changing in a chaotic fashion. This appears to be a result of the way the tidal pull from Saturn varies on account of Hyperion's eccentric orbit and elongated shape.

Phoebe is Saturn's outermost substantial satellite, and is the innermost of those travelling in retrograde orbits. It has a rotation period of about nine hours (less than a thousandth of its orbital period) and is the only satellite of Saturn other than Hyperion not to be in synchronous rotation. Phoebe is almost certainly a captured asteroid. It lacks the degree of redness associated with most asteroids beyond 3.4 AU, and the darkness of its surface (albedo 0.06) suggests that it belongs to the carbonaceous variety. Theoretically, black dust thrown off as crater ejecta from Phoebe should spiral in towards Saturn. Much of this dust would be swept up by the leading hemisphere of the next satellite in from Phoebe, Iapetus, which of course orbits in the opposite direction. Iapetus does indeed have a darkened leading hemisphere but, as discussed later, its colour does not match that of Phoebe. Phoebe was imaged by Cassini on its initial approach to Saturn, revealing a battered surface (Figure 10.7). Bright crater-walls appear to show an icy interior that is elsewhere buried by about 500m of darker dusty material.

Based on their orbital characteristics, Saturn's small irregular outer satellites belong to several groups. As suggested for Jupiter, each such group is probably the remains of a larger body that broke up during capture. The seven regular satellites of Saturn large enough to be virtually spherical in shape are discussed in turn below, beginning closest to Saturn and working outwards. All except Titan are low density, icy bodies lacking substantial atmospheres, and these were imaged moderately well by the two Voyager probes, though we lack detailed global coverage.

Mimas and Enceladus

	Mimas	Enceladus
Equatorial radius (km)	199	249
Mass (relative to Earth)	0.0000062	0.000011
Density (g/cm³)	1.14	1.00
Surface gravity (relative to Earth)	0.0063	0.0072
Composition of surface	ice	ice
Mean surface temperature	−185°C	−185°C

table 10.3 planetary facts for Mimas and Enceladus

The two innermost of Saturn's regular satellites are Mimas and Enceladus. They resemble each other in basic physical properties (Table 10.3), but are remarkably different in their histories. Mimas is surfaced by fairly clean ice and is a heavily cratered little world (Figure 10.8). However, there are few craters more than 30 km across, with the notable exception of a 130 km crater named Herschel after Sir William Herschel who discovered both Mimas and Enceladus in 1789. The crater Herschel is so big that if the impacting body responsible for forming it had been travelling faster or had been slightly more massive, the force of the impact would probably have been sufficient to break Mimas into fragments.

A satellite such as Mimas that orbits relatively close to a giant planet is especially prone to powerful impacts, because the planet's gravity both focuses and accelerates debris approaching the planet from elsewhere in the Solar System. For all we know, Mimas might actually have been broken apart by a previous

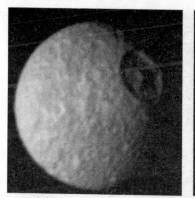

figure 10.8 Voyager 1 images of two sides of Mimas (398 km diameter) the view on the left shows the crater Herschel, with its obvious central peak

impact. If so, the satellite we see today would have re-accreted from these fragments, and later have been lucky to escape re-fragmentation by the Herschel-forming impact. Irrespective of this possibility, counting the number of craters on such a body is no good for telling us the absolute age of the surface. We have no way of determining either the general rate of cratering in the Saturn system, or of allowing for debris-creating events such as fragmentation of a neighbouring satellite by a rogue impact which would have caused a flurry in the local rate of cratering.

However, so far as we can tell from the Voyager images, which show details as small as 2 km in some parts of Mimas, the current surface has been shaped passively by impact cratering and there is no sign of internal activity. The situation could hardly be more different on Enceladus (Figure 10.9). Here is a world scarcely bigger than its neighbour Mimas, but which shows abundant signs of a prolonged active geological history.

The crater density on Enceladus varies greatly from region to region, and even the most densely cratered parts of Enceladus have about the same density of craters as the least cratered areas on Saturn's other satellites. Some regions are apparently craterless and smooth, but other craterless regions are traversed by sets of curved parallel ridges. In some places craters are truncated by terrain boundaries, so that part of a crater has survived although the rest has been obliterated by whatever process created the adjacent younger surface.

Clearly Enceladus has experienced several episodes of activity, presumably driven by tidal heating. Enceladus is in 2:1 orbital resonance with the next-but-one satellite, Dione. At present Enceladus's orbit is so nearly circular that the rate of tidal heat

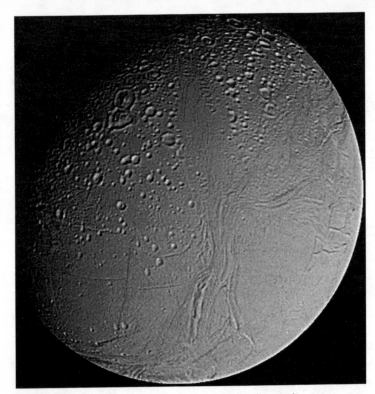

figure 10.9 Voyager 2 view of Enceladus (498 km diameter)
note the contrast between heavily cratered areas, weakly cratered areas, and
tracts of both smooth and ridged craterless terrain

production is probably too slight to drive activity. However, if Enceladus's orbit has been more eccentric at various times, perhaps through interaction with Janus and Epimetheus, which are in 2:1 orbital resonance with it, then this could account for the degree of tidal heating necessary to explain the extent of the resurfacing.

What went on at the surface is clear in a general sense, but the details must await improved imaging which may also indicate whether activity continues on Enceladus today or has turned off (presumably only temporarily). The sharp edges between some of the terrain units appear to be fractures or faults, and the resurfacing must have been achieved by flooding by some kind of cryovolcanic lava. The ice in Saturn's satellites is likely to be contaminated not just by salts, as in Jupiter's icy galilean satellites, but also by ammonia. This substance might have been able to condense from the solar nebula at the radius of Saturn's orbit and certainly should be present at Uranus and beyond. An ice formed from a water–ammonia mixture could begin to melt at –97 °C, requiring even less heating than the salty or sulfuric acid-rich melts suggested for Europa. Ammonia has not yet been detected spectroscopically on any icy satellite, which is thought to be because exposure on the surface to the vacuum of space in a harsh radiation environment breaks it down in a relatively short timescale. However, the Cassini mission should be able to detect any ammonia still present in freshly emplaced or freshly exposed surfaces.

Another remarkable feature of Enceladus is that it has a brighter surface than any other icy satellite, with an albedo of virtually 1. This is not simply a consequence of the extreme youth of the craterless parts of its surface, because it is bright all over. The surface may be covered by a fresh powder of frost distributed globally by an explosive cryovolcanic eruption. Alternatively, it could reflect the fact that Enceladus's orbit runs along the densest part of Saturn's E ring, which is likely to be composed of smaller icy particles than the main rings. It was once thought that the E ring itself must indicate recent eruptions on Enceladus, because this ring is theoretically an ephemeral feature as the particles within it are only about a micrometre across and therefore vulnerable to dispersion by radiation pressure and Saturn's magnetic field. However, it is now believed that once such a ring has formed it could be self-sustaining, with the rate of loss of particles being balanced by creation of new ones through impacts of ring particles onto Enceladus, some of the micro-ejecta from which would resupply the ring.

Tethys and Dione

	Tethys	Dione	Rhea	Iapetus
Equatorial radius (km)	529	560	764	720
Mass (relative to Earth)	0.00010	0.00018	0.00036	0.00027
Density (g/cm³)	1.00	1.50	1.24	1.0
Surface gravity (relative to Earth)	0.015	0.024	0.027	0.021
Composition of surface	ice	ice	ice	dusty ice
Mean surface temperature	–185 °C	–185 °C	–185 °C	–178 °C

table 10.4 planetary facts for Tethys, Dione, Rhea and Iapetus

Tethys and Dione constitute another size pair, and are even more similar to one another than is the case for Mimas and Enceladus. Their physical properties are summarized along with two other mid-sized icy satellites of Saturn in Table 10.4. Neither has any tracts of particularly young surface, but on both the crater density varies from place to place. Furthermore, both are traversed by fracture systems indicating that these have not just been Mimas-like passive ice-balls throughout their existence.

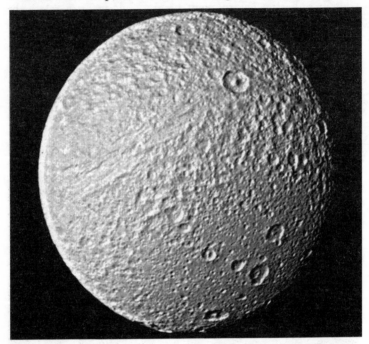

figure 10.10 Voyager 2 view of the Saturn-facing hemisphere of Tethys (1058 km diameter)

the largest crater (upper right, 100 km diameter) still has its original raised rim, and is on an old fairly heavily cratered surface

however, the less cratered (and therefore younger) plains in the lower right have been resurfaced by icy lava that buried all the pre-existing craters except for the five largest, which were submerged up to their rims

the fracture system extending towards the lower left from the 100 km crater is associated with the largest impact on Tethys, the 440 km crater Odysseus which lies on the opposite hemisphere

Tethys has a particularly low density. The small fraction of internal rock and hence weaker radiogenic heating that this implies may account for the fact that Tethys shows fewer signs of past activity than Dione. However, a clear regional variation in crater density is revealed on Tethys in Figure 10.10. Most of the hemisphere seen in this view is densely cratered, by impacts of a wide range of sizes that would be consistent with sweeping up of the debris left over at the end of planet formation. By contrast, in the lower right of this hemisphere there are significantly fewer craters and their size range is limited. They tend not to touch or overlap as elsewhere, and the intervening ground appears smooth and flat. This ground was presumably formed by cryovolcanic flooding from unidentified eruption sites. The five largest craters in this region (four of which have noticeable central peaks) are remarkable in that the level of the surrounding plain is almost up to their external rims. These must be pre-existing craters whose ramparts were just high enough to prevent their interiors being inundated by the icy lavas that buried their smaller neighbours.

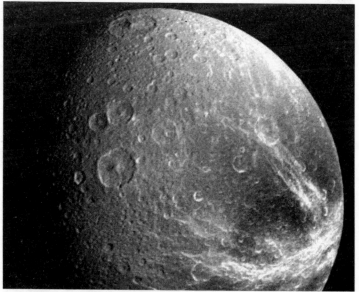

figure 10.11 Voyager 1 image of part of Dione's Saturn-facing hemisphere the largest crater, Aeneas, seen foreshortened near the top, is about 160 km across

The craters we see now on the plains are younger, and formed after the flooding event. The more restricted size range of these craters, as compared to those on the older terrain, suggests that most of these craters are a result of impacts of debris originating within the Saturn system. Of course, there must be craters of the same age and origin on the older terrain too, but they are not so easy to spot because they are surrounded by older craters that have not been buried by flooding.

Dione has a wider range of crater densities than Tethys. This is not readily apparent on the image shown in Figure 10.11, but this view does show Dione's characteristic pale wispy markings that appear to be associated with faulted troughs that cut many parts of the globe. To judge from regional variations in crater density, there have been at least two episodes of local cryovolcanic flooding, widely spaced in time, that have affected different regions. These episodes may have been times when Dione was tidally heated by interaction with Enceladus. If so, they happened long before the tidal heating events responsible for the resurfacing we see on Enceladus today, because Dione is far more cratered than Enceladus. On the other hand, even the oldest cratered terrain on Dione has fewer craters, particularly those greater than 20 km across, than the least densely cratered areas seen on Rhea, which is discussed next.

Rhea

The Voyager image coverage of Rhea was frustratingly incomplete. Details as small as 2 km were revealed in the north polar region, which is heavily cratered and looks how all icy satellites were expected to be before the role of episodic tidal heating was recognized (Figure 10.12). However, images of most of the Saturn-facing hemisphere are only good enough to show details down to 20 km, and on the opposite hemisphere the spatial resolution of the best images is even poorer. So far as can be told there are a few fractures and some variations in crater density, but Rhea has little in common with Dione and Tethys other than the fact that Hubble Space Telescope observations have discovered ozone trapped within the ice on both Rhea and Dione. This presumably formed by breakdown of water molecules in the same manner as on Jupiter's icy galilean satellites.

figure 10.12 Voyager 1 image of part of Rhea's heavily cratered north polar region (about 600 km from top to bottom), which was the area best seen

Titan

Equatorial radius (km)	2575
Mass (relative to Earth)	0.022
Density (g/cm³)	1.88
Surface gravity (relative to Earth)	0.14
Composition of surface	ice and hydrocarbons
Mean surface temperature	−180 °C
Composition of atmosphere	nitrogen (82–99%), methane (1–6%), argon (1–6%), hydrogen (0.2%), carbon monoxide (0.005%), ethane (0.002%)
Atmospheric pressure at surface (relative to Earth)	1.5

table 10.5 planetary facts for Titan

Of all the planetary satellites in the Solar System, Titan is the only one with a substantial atmosphere. This makes Titan a particularly intriguing place, and it is the first outer planet satellite to have been targetted by a landing probe. This was the Huygens lander, which detached from the Cassini Saturn orbiter and made a successful parachute descent to Titan's surface in January 2005 (Table 10.1). Cassini mapped Titan's surface by imaging radar, and Huygens carried optical imagers for use during descent and a variety of analytical experiments to examine the nature of the atmosphere and surface. Voyager images of Titan show only an opaque orange atmosphere overlain by a blue haze layer, so until Cassini-Huygens arrived our knowledge of Titan's surface was limited to what can be deduced from Earth-based observations.

Titan has a similar density to Ganymede and is almost as large (Table 10.5). Indeed, some old tables augment Titan's size by the 200 km thickness of the opaque part of its atmosphere, on which dubious basis Titan becomes the largest planetary satellite of all. Clearly Titan's density argues for it being largely an icy body, but we do not know to what extent it has become internally differentiated to form a rocky or iron-rich core.

Titan's nitrogen- and methane-rich atmosphere is inherited from the gases scavenged directly from the solar nebula and the circum-Saturn gas and dust cloud within which Titan grew, supplemented by a probably small degree of subsequent degassing from Titan's interior. The reason why the young Titan was able to scavenge and retain an atmosphere whereas Ganymede and Callisto did not is that Titan grew in a colder environment, being both further from the Sun and receiving less warmth from its planet. Saturn's other satellites are much smaller, so their gravity is insufficient to cling on to any gas.

Importantly, Titan's atmosphere shields its surface from radiation, so water-ice does not get broken down or liberated as a vapour, and of course it is far too cold for ice to evaporate naturally. This means that Titan's atmosphere is both dry and lacks oxygen. As a result, it contains compounds far rarer in the present atmospheres of the terrestrial planets, but which may have been abundant there too originally. Those detected spectroscopically include hydrocarbons such as ethane (C_2H_6), ethene (C_2H_4), ethyne (C_2H_2) and propane (C_3H_8) and nitrogen compounds such as hydrogen cyanide (HCN), cyanogen (C_2N_2) and cyanoacetylene (HC_3N). Many of these gases condense in Titan's atmosphere to contribute to the orange smog that

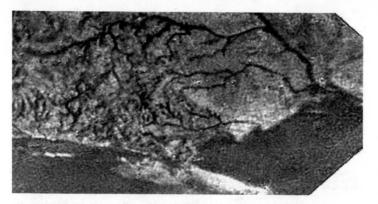

figure 10.13 an 8 km wide portion of the surface of Titan, imaged by
Huygens from below the cloud layer during its parachute descent
bright high ground is cut by valleys that drain towards the dark flat area in
the lower part of the image, which may be either a sea or a swamp coated by
hydrocarbon sludge
the very bright features near the bottom of the image are low cloud or fog

obscures the surface. However, the colour is probably caused by
the linking together of hydrocarbon molecules into longer
chains than those listed above, because of reactions triggered by
solar ultraviolet radiation. Similar 'photochemical' reactions
could have produced Titan's ethane from methane.

Ethane could play an especially fascinating role in Titan's
meteorology, because condensed droplets of ethane ought to be
able to fall to the surface and collect there as liquid. In other
words, Titan is probably a cold, dark and gloomy place made
yet more miserable by a continual drizzle of ethane. Winds blow
at 100 m per second in Titan's upper atmosphere in the direction
of Titan's rotation, but we have no knowledge of the near-
surface wind pattern.

The strength of radar reflections obtained using ground-based
radars showed that Titan is not covered by a global ocean of
ethane as was once suggested, but did not rule out seas and
lakes from which ethane could evaporate back into the
atmosphere to complete the cycle. The Huygens probe was
designed to float should it touch down in one of these bodies of
liquid, and to measure its density and depth.

In fact Huygens landed on a solid surface, strewn with pebbles
of ice that were probably eroded from nearby highlands and

transported by rivers of ethane or methane. A valley network was spectacularly imaged during Huygens' descent (figure 10.13). Titan's landforms are thus strikingly Earth-like, with erosional features such as valleys and cliffs and depositional features such as flood-plains and beaches. Add to this an intricate weathering cycle involving hydrocarbons mixed with or coating the surface ice, and the possibility of transport and re-working of surface material by the wind, and it is clear that Cassini-Huygens has only begin to reveal the wonders awaiting discovery on Titan's surface.

Iapetus

Iapetus is the outermost of the regular satellites of Saturn. Its comparatively large size and great distance from the glare of the planet resulted in its being the second of Saturn's satellites to be discovered, by Giovanni Cassini in 1671. Iapetus has the remarkable property of appearing much brighter when lying to the west of Saturn in the sky (when we see its trailing hemisphere) than when lying to Saturn's east (when we see its leading hemisphere). This is because Iapetus's trailing hemisphere reflects about 50 per cent of the incident sunlight (a value typical of most icy satellites) but the leading hemisphere reflects less than 10 per cent. As mentioned previously, this darkening could be caused by dust from the innermost retrograde satellite Phoebe. However, Iapetus's leading hemisphere is red (possibly indicating the presence of tarry tholins) as well as being dark, and so in that respect it appears to differ from Phoebe's composition.

The best Voyager image of Iapetus is shown in Figure 10.14. This shows that the edge of the dark zone is sharp rather than diffuse, and irregular rather than smooth in outline. Furthermore, the floors of some of the craters near the edge of the bright hemisphere are covered by dark material. These observations are not consistent with the dark material being simple, unmodified fall-out from Phoebe, and at the least there must be some internally driven process at work to explain these characteristics. However, an ancient and passive surface is indicated by the heavy degree of cratering, which is apparent globally except in the dark hemisphere which was too dark to reveal any such features on Voyager images. Iapetus is another place where the improved spectral data and images from the Cassini orbiter may well lead to a body's nature and history being dramatically reassessed.

figure 10.14 the best Voyager image of Iapetus, showing part of the dark leading hemisphere on the left and the brighter trailing hemisphere on the right

Uranus

In this chapter you will learn:
- about Uranus, a giant planet tipped on its side. It has a complex ring system and a large and varied family of satellites.

Planetary facts	
Equatorial radius (km)	25,559
Mass (relative to Earth)	14.54
Density (g/cm³)	1.32
Surface gravity (relative to Earth)	0.89
Rotation period	17.24 hours
Axial inclination	97.9°
Distance from Sun (AU)	19.22
Orbital period	83.75 years
Orbital eccentricity	0.046
Composition of surface	gassy
Mean cloud-top temperature	−214 °C
Composition of atmosphere	hydrogen (83%), helium (15%), methane (2%)
Number of satellites	27

Uranus was the first planet to be discovered since antiquity. It is barely bright enough to be visible to the unaided eye, so it is not surprising that it escaped notice until the telescopic era. It was Sir William Herschel who first happened upon Uranus, in 1781, and six years later he discovered its two outermost substantial satellites Oberon and Titania. Uranus is a giant planet, but like Neptune it lacks the enormously thick blanket of hydrogen possessed by Jupiter and Saturn, despite having a similar sized icy and rocky core (Figure 2.5).

Rotation and orbit

Uranus creeps round the Sun at more than twice the distance of Saturn and at less than 7 km per second. As a result it is overtaken by the Earth every 370 days. Its internal rotation period is the slowest of the four giant planets (but still faster than the Earth's), and the globe is flattened only to the extent that its polar radius is about 2.3 per cent less than its equatorial radius.

The most remarkable aspect of Uranus's rotation is that its axial inclination is nearly 98°. Technically this means that Uranus rotates in a retrograde direction, but it is simplest to think of the planet as tilted over onto its side and slightly overturned. This peculiar orientation means that for almost half of each orbit one pole never sees the Sun, and for substantial periods the sunlit pole receives more solar illumination than the equator.

How Uranus came to be on its side is a mystery. In view of the consistency in alignment of the rotational axes of the other major planets, it is unlikely that Uranus could have inherited its orientation directly from the solar nebula. It is possible that Uranus was knocked over by a particularly severe giant impact late in its formation. This would have had catastrophic consequences for any satellites that were already in existence. If not destroyed by debris from the impact, they would have subsequently been vulnerable to tidal disruption and mutual collisions while tidal forces were re-orienting their orbits into the new equatorial plane.

Missions to Uranus

There has been only one mission to Uranus, Voyager 2 which flew past in January 1986. At that time the planet's south pole was pointed approximately towards both the Sun and the direction from which Voyager 2 approached. Consequently, Uranus and the orbits of its satellites were oriented like a bull's eye target with respect to the incoming spaceprobe. This meant that the probe's trajectory could not be arranged so as to pass by each satellite plus the planet in turn, as was possible at Jupiter and Saturn where the orbits all lay close to the ecliptic plane. Instead, the choice was made to pass close to the innermost previously known satellite, Miranda, and perform a gravity assist manoeuvre close to Uranus in order to proceed onwards for Voyager's final encounter with the Neptune system.

The interior

Uranus has virtually the same density as Jupiter. However, this does not mean that its composition is the same, because Jupiter's enormous mass results in much stronger gravity and a correspondingly greater amount of compression in its interior. The pressure within Uranus is not sufficient to form metallic hydrogen, and indeed scavenging of gas from the solar nebula by the proto-Uranus appears to have been terminated before more than about two Earth-masses of hydrogen had been collected around its core. The rocky central core is probably at a pressure of about 8 million atmospheres and a temperature of around 8000 °C.

Alone among the giant planets, Uranus does not radiate significantly more energy than it receives, which tells us that heat is not being produced internally through contraction or differentiation. However, it does have a powerful magnetic field, about 50 times the strength of the Earth's. This is inclined at 59° to the planet's rotational axis and provides a convenient way to determine the internal rotation speed. The centre of symmetry of Uranus's magnetic field is roughly one-third of the way from the middle of the planet to the surface. This strongly suggests that the field is produced by convection currents in a shell within the icy part of the planet's interior. Uranus's 'ice' is likely to contain a substantial proportion of methane and ammonia in addition to water, and is probably predominantly liquid rather than solid.

The atmosphere

With its south pole in full sunlight at the time of the 1986 Voyager 2 fly-by and subsequently during study using the Hubble Space Telescope, the pattern of atmospheric circulation on Uranus differs from that seen on Jupiter and Saturn. Warmed air rises at high southern latitudes and flows equatorwards where it sinks. During its journey from the sunlit pole towards the equator the air lags progressively further behind the planet's rotation. As a result, by the time it reaches the equator the prevailing wind speed is about 100 m per second towards the west, in the opposite direction to the equatorial winds of Jupiter and Saturn.

Uranus's visible cloud tops are methane, which does not condense in the warmer atmospheres of Jupiter and Saturn. There may be unseen ammonia clouds at a deeper level. The overall colour of Uranus is green, because of absorption of red light by the otherwise transparent methane gas that overlies the cloud deck.

Compared to the other giant planets, Uranus's atmosphere is rather featureless and contained no storm systems when Voyager 2 flew past (Figure 11.1). However, images by the Hubble Space Telescope in 1994 revealed high altitude clouds in the southern hemisphere (Figure 11.2). By 1997 bright clouds were visible to the north of the equator, at latitudes that had been hidden in darkness during the Voyager 2 fly-by.

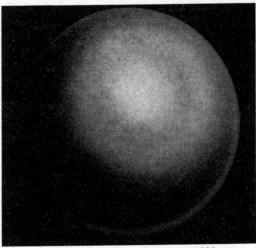

figure 11.1 Voyager 2 image of Uranus in January 1986
the south polar region appears brightest because the cloud layer is at a higher altitude there

Rings

Uranus's rings were discovered in 1977 by stellar **occultation**, in other words noting the dimming of the light from a star as the rings, the planet and then the rings again passed in front of it. Uranus was thus the second planet, after Saturn, known to have a ring system. Jupiter's rings were not confirmed until two years later, when Voyager 1 got there in 1979. Most of our knowledge of Uranus's rings comes from the 1986 Voyager 2 fly-by (Figure 11.3) and subsequent Hubble Space Telescope studies (Figure 11.2). These show that Uranus has ten slender rings, mostly composed of dark boulders (10 cm to 10 m in size with an albedo of about 0.015), but that there is diffuse dust between the main rings (Figure 11.4). There is also a 3000-km-wide ring of diffuse dust extending towards the planet beginning 1500 km inside the innermost of the narrow rings that is reminiscent of Saturn's D ring. The total mass of Uranus's rings is more than that of Jupiter's but less than that of Saturn's.

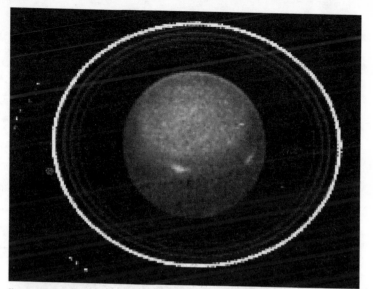

figure 11.2   Hubble Space Telescope image of Uranus in August 1997
note the bright hood over the south pole and the two white high-altitude clouds
this image also shows four of the planet's rings, which appear wider than
they truly are because of the limited resolution of the imaging system
the motion of three of Uranus's tiny inner satellites has been revealed by
superimposing a set of three images recorded six minutes apart, which
shows each satellite as a string of three dots
clockwise from the bottom these are Portia, Juliet and Cressida

The narrow rings are typically less than about 10 km in width,
and are remarkable because at least six are inclined by a fraction
of a degree to the planet's orbital plane. This contrasts with the
apparently exact alignment of other planetary rings, and furthermore several of Uranus's rings are variable in width and not quite
circular in shape. The outermost ring, known as the epsilon ring,
is the most substantial and is visible in Figures 11.2 to 11.4. This
is at a mean distance of 51,150 km from the planet's centre but
is eccentric in shape so that its distance from the planet varies by
800 km around its circumference. Its width varies between 20
km where closest to Uranus and 100 km where furthest from it.
As Figure 11.3 shows, this ring is kept in place by two shepherd
satellites (Ophelia and Cordelia) that perform the same role as
Prometheus and Pandora for Saturn's F ring, though without (so
far as we know) twisting it into a braid.

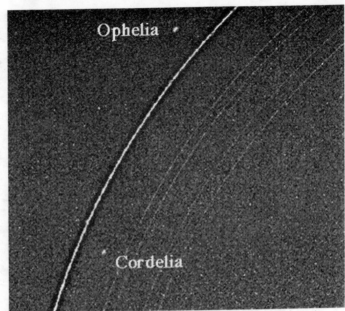

figure 11.3 Voyager 2 image showing Uranus's outermost and most prominent ring, the epsilon ring, with its two shepherd satellites Ophelia and Cordelia

within the orbit of Cordelia it is possible to make out the fainter delta, gamma, beta and alpha rings

unlike Figure 11.2, this image shows the rings at their true widths and satellites at their true sizes

figure 11.4 Voyager 2 view looking towards the Sun through Uranus's rings, revealing dust between the slender boulder rings

this is a 96-second exposure, and the stars seen beyond the rings appear as streaks because of the spacecraft motion

Satellites

Name	Distance from planet's centre (km)	Radius (km)	Orbital period (days)	Mass	Density (g/cm³)
Cordelia	49,752	13	0.335	?	?
Ophelia	53,763	16	0.376	?	?
Bianca	59,166	22	0.435	?	?
Cressida	61,767	23	0.464	?	?
Desdemona	62,658	29	0.474	?	?
Juliet	64,378	42	0.493	?	?
Portia	66,097	42	0.513	?	?
Rosalind	69,927	29	0.558	?	?
S/2003 U2	74,800	12	0.618	?	?
Belinda	75,256	34	0.624	?	?
S/1986 U10	76,160	40	0.638	?	?
Puck	86,004	77	0.762	?	?
S/2003 U1	97,700	16	0.923	?	?
Miranda	129,800	240 × 234 × 233	1.413	6.59×10^{19} kg	1.20
Ariel	191,240	579	2.520	1.35×10^{21} kg	1.67
Umbriel	266,000	585	4.144	1.17×10^{21} kg	1.40
Titania	435,840	790	8.706	3.53×10^{21} kg	1.71
Oberon	582,600	760	13.46	3.01×10^{21} kg	1.63
S/2001 U3	4,280,000	6	799 R	?	?
Caliban	7,231,000	30	580 R	?	?
Stephano	8,004,000	10	677 R	?	?
Trinulo	8,571,000	< 10	758 R	?	?
Sycorax	12,179,000	95	1288 R	?	?
S/2003 U3	14,345,000	6	1695	?	?
Prospero	16,234,000	15	1977 R	?	?
Setebos	17,501,000	15	2235 R	?	?
S/2001 U2	21,000,000	6	2823 R	?	?

table 11.1   satellites of Uranus
S/1986 U10 is the provisional designation of a contentious satellite, as
described in the text (R = retrograde, ? = value unknown)

Uranus's family of satellites (Table 11.1) conforms to the norm
for giant planets, in that it has an inner set of a dozen irregular-
shaped moonlets (all of them unknown until discovered on
Voyager 2 images or recent telescopic survey). As shown in
Figure 11.3 Cordelia and Ophelia are shepherd satellites either

side of the epsilon ring, and the other inner moonlets orbit not far beyond this ring (e.g. Figure 11.2). Only the largest and outermost of these, Puck, was imaged well enough by Voyager to reveal its shape and surface features (Figure 11.5). This is irregular but not elongated in shape and several craters are hinted at in the image. In appearance Puck is reminiscent of Saturn's moonlets Janus and Epimetheus (Figure 10.6), but whereas those are bright icy objects with albedos of 0.8–0.9, Puck is a dark body with an albedo of only 0.07. This is not apparent when comparing the images of these bodies, because they have been processed differently in order to reveal the surface features to best advantage.

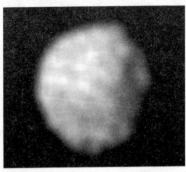

figure 11.5 the best Voyager 2 image of Puck (154 km across) features as small as 10 km across can be made out, the most obvious of which is a 40-km-diameter crater on the right

Uranus's other inner satellites appear to be dark like Puck, and have all been given the name of a female Shakespearean character. Those composed of material that collected from gas and dust around Uranus would be expected to include contaminants such as ammonia, methane and carbon monoxide within their ice. If this is so, their low albedos could be a result of radiation-darkening of methane with age, either because of solar ultraviolet radiation or because of charged particles channelled by Uranus's magnetic field. Either form of radiation could knock hydrogen atoms out of methane molecules, ending up either with methane being converted to carbon, or else damaged methane molecules linking into chains to create tholins.

In 1999 an 11th inner moonlet of Uranus was noticed on a series of 13-year-old Voyager images, sharing almost the same orbit as Belinda. This was given the provisional designation S/1986 U10, indicating that it was the tenth probable satellite of Uranus to have been discovered on 1986 observations (Puck, being bigger than the rest, was detected in 1985 before Voyager 2 reached Uranus). However in 2001 the International Astronomical Union decreed that S/1986 U10 must no longer be regarded as a proven satellite because two years had passed since the announcement of discovery without it having been seen by telescope. It is real though, having been rediscovered using the Hubble Space Telescope in 2003.

Beyond its inner moonlets, Uranus has five sizeable regular satellites whose surfaces are markedly brighter (albedos 0.19–0.35) and which are discussed individually below. Beyond these in turn are five small irregular satellites discovered telescopically in 1997 and 1999, with names taken from Shakespeare's play *The Tempest*, and a sixth as yet unnamed, discovered in 2001. These all have inclined retrograde orbits, and are almost certainly captured objects. Their shapes are unknown, and their sizes listed in Table 11.1 are guesses based on the assumption that each has an albedo of less than 0.1. With eccentricities of 0.588 and 0.522 respectively, the orbits of Setebos and Sycorax are more elongated than that of any other planetary satellite except Neptune's outermost large satellite, Nereid.

Miranda

	Miranda	Ariel	Umbriel	Titania	Oberon
Equatorial radius (km)	236	579	585	790	760
Mass (relative to Earth)	0.000011	0.00023	0.00020	0.00059	0.00050
Density (g/cm³)	1.20	1.67	1.40	1.71	1.63
Surface gravity (relative to Earth)	0.008	0.027	0.023	0.039	0.036
Composition of surface	ice	ice	ice	ice	ice
Mean surface temperature	−200 °C	−200 °C	−200 °C	−200 °C	−200 °C

table 11.2   planetary facts for the largest satellites of Uranus

It was a happy circumstance that led to the trajectory chosen for Voyager 2 passing close to Miranda, because this turned out, quite unexpectedly, to be a complex and fascinating little world (Figures 11.6 and 11.7). It is on the size limit for an icy body to be pulled into a virtually spherical shape by its own gravity. Not being in orbital resonance with its neighbours, it was predicted to be a passive ice ball like Mimas, but the reality is far more interesting.

figure 11.6 mosaic of Voyager 2 images of Miranda (572 km diameter) the three visible coronae are named Arden (lower left), Inverness (lower centre) and Elsinore (upper right)

Miranda's trailing hemisphere is on the right and its leading hemisphere is on the left

Like all Uranus's satellites seen by Voyager 2, only the southern hemisphere was sunlit, so we only have knowledge of half of the globe. What we see is quite unlike any other known planetary body. Just over half the area seen is heavily cratered, but most of the craters are blurred, as if blanketed with dust or 'snow', perhaps dispersed from explosive cryovolcanic eruptions. The rest of the visible hemisphere is occupied by three remarkable tracts of terrain, each one referred to as a corona though they do not closely resemble each other and have little in common with the coronae on Venus other than a vaguely concentric pattern. They are named Arden, Inverness and Elsinore after the settings of some of Shakespeare's plays. Craters on these coronae are fresh and sharp in appearance, as are the younger craters elsewhere, so evidently the global mantling of the older terrain took place before the coronae were created.

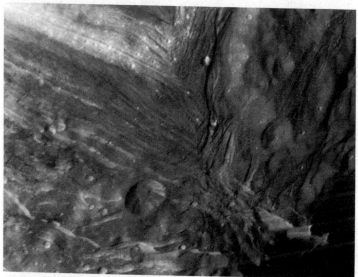

figure 11.7 part of the most detailed image of Miranda, showing details as small as 500 m across, which is the best spatial resolution of any Voyager image

the area covered is about 80 km across and part of Inverness corona is seen at the upper left

the cliff at the lower right is over 10 km high, but it slopes at about 60 degrees rather than being vertical

this and several fractures parallel to it are geological faults that run up the left-hand side of Inverness corona in the orientation seen on Figure 11.6

This much is straightforward, but what actually generated the coronae is unclear. Frustratingly, Arden and Elsinore coronae are only partly covered by the images, so their complete shape is unknown. When the images were first received it was suggested that Miranda is a body that re-accreted following collisional break up. Each corona was thought to represent a discrete fragment, with the tonal banding apparent in Arden and Inverness coronae representing cross-sections through the interior of the original body. This now seems unlikely, if only because of the much greater cratering age of the intervening terrain. The edges of Inverness corona appear to be related to faults (Figure 11.7), but this and other coronae may overlie internal density anomalies inherited from a re-accretion event. If so, each corona could be a result of volcanic eruptions triggered by a past episode of tidal heating through orbital interaction with Ariel or Umbriel. Possibly, the bright patches on Arden and Inverness coronae are fresh powder erupted from explosive vents. On the other hand Elsinore corona lacks tonal variations and is more likely to be a product of multiple eruptions of cryovolcanic lava. In contrast with Saturn's satellites, where the presence of ammonia is possible but not required by models of Solar System formation, it is extremely likely that Uranus's icy satellites contain a significant proportion of ammonia mixed with the water-ice. The lavas this is likely to produce would be highly viscous, and this could explain Elsinore corona's bulbous ridges.

Ariel

The second-best imaged of Uranus's satellites is Ariel (Figure 11.8). Ariel is denser than any of Saturn's satellites except the much larger Titan, and rock must make up about half its mass. This is a complex globe, whose oldest cratered terrain is cut by numerous faults that divide the cratered terrain into blocks. Unlike the situation on Miranda, none of Ariel's craters have had their topography muted by a blanket of superficial deposits. Ariel's cratered terrain is markedly less cratered, and therefore younger, than the most heavily cratered regions on Uranus's other large satellites. Many of the faulted valleys separating Ariel's blocks of cratered terrain are floored by smoother and less cratered (and therefore younger) material, which is almost certainly an ammonia-rich cryovolcanic lava. This is especially notable in the lower part of Figure 11.8, where in some places it has spilled out beyond the ends of valleys and flooded the adjacent areas.

If Ariel's interior is rich in ammonia, the effect of this in allowing ice to melt at very low temperatures may mean that radiogenic heat production was sufficient to liberate ammonia-water melts for the first 2 billion years of Ariel's history. This would have led to complete differentiation by way of forming a silicate-rich core, and could coincide with the epoch when Ariel's cratered terrain was formed. What may have initiated the fracturing of this oldest surviving surface might have been freezing of a previously molten deep part of its icy mantle. Ices are generally less dense than their equivalent liquids, and the change in volume caused by expansion during freezing would increase Ariel's radius by about a kilometre. This could have

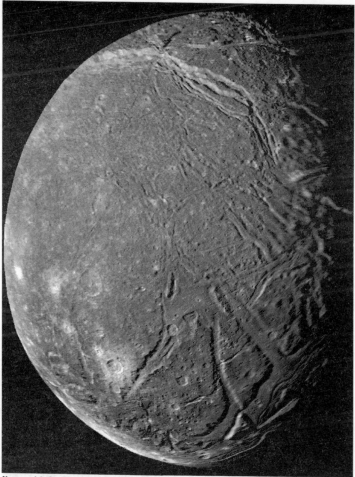

figure 11.8 Voyager 2 view of Ariel (1158 km diameter)
Ariel's south pole is towards the left-hand edge of the disk and its trailing
hemisphere is towards the top

stretched and fractured the lithosphere, thereby initiating the
faulted valleys we see today. Most are probably extensional
features, though there are intriguing hints of sideways
movement across some of them. Cryovolcanic magmas could
have reached the surface via these fractures almost immediately.
Other batches of magma could have escaped subsequently when
additional melts were produced by tidal heating during episodes
of orbital resonance with Umbriel.

Another, but possibly not unrelated, theory for the origin of Ariel's global fracture pattern is that it was controlled by tidal stresses. It is frustrating that 65 per cent of the globe remains unimaged, because if we could see more of Ariel's fascinating surface undoubtedly this would trigger new ideas about its geological history and the forces that have shaped it.

Umbriel

Umbriel is the next satellite out from Ariel. It is Ariel's near-twin in size and only slightly less dense, therefore it is curious that Umbriel appears so different. Perhaps similarities would be revealed if we had images of the unseen two-thirds of each globe.

As its name suggests, Umbriel is darker than the other large satellites of Uranus, with an albedo of 0.19. The imaged part of the globe is much more heavily cratered than the cratered terrain on Ariel, and comparable with the lunar highlands. Umbriel's surface is therefore old, and its dark colour could be a result of exceptional radiation-darkening of methane with age or a particularly high carbon content.

The only bright material visible on Umbriel is in the central peak of the 110 km crater Vuver and the floor of the 150 km crater Wunda (Figure 11.9), which have albedos of about 0.5. Vuver's central peak could be bright because it exposes material uplifted from a depth of about 10 km that differs in composition from the normal surface material. Wunda's floor could be covered by a pale variety of cryovolcanic lava, which might represent the most recent activity in the imaged part of the globe.

Titania

The remaining two satellites, Titania and Oberon, are larger than Ariel and Umbriel, but similar in size to each other. Titania and Oberon are named after the fairy queen and king in Shakespeare's *A Midsummer Night's Dream*. In that play, appropriately enough for characters whose names were later given to moons, Oberon famously greets Titania with the words, 'Ill met by moonlight, proud Titania'. They clearly did not get on well, so it is appropriate that in their surface features Titania is more like Ariel whereas Oberon is more like Umbriel.

The best image of Titania is only good enough to show details down to about 7 km across (Figure 11.10). This is adequate to reveal a surface that is more heavily cratered than Ariel but less

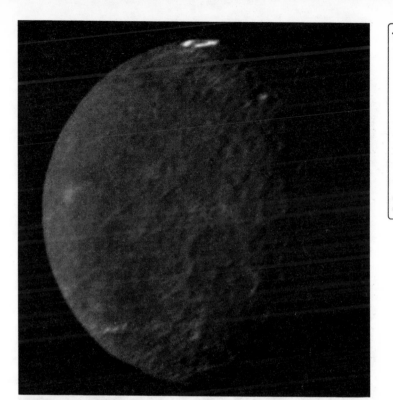

figure 11.9   the best Voyager 2 image of Umbriel (1170 km diameter)
the bright-floored crater Wunda seen foreshortened near the top lies near the
middle of the trailing hemisphere
the crater Vuver, remarkable for its bright central peak, lies below and to
the right

cratered than Umbriel or Oberon, and which is traversed by
fault scarps between 2 and 5 km high and up to 1500 km long.
Maybe these have the same origin as the faults on Ariel, but
there are no visible signs that Titania's faulted valleys have been
flooded by cryovolcanism. Another indication that Titania has
been active more recently than Umbriel and Oberon is that the
topographic expression of its largest craters is muted, as if the
lithosphere at the time they were formed did not have sufficient
strength to support such large features.

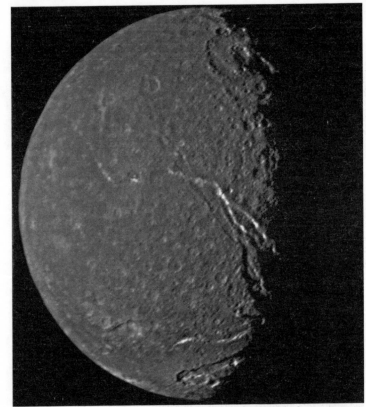

figure 11.10   the best Voyager 2 image of Titania (1580 km diameter)
the trailing hemisphere is towards the top
note the crater near the bottom that is cut by a faulted valley, known as
Belmont chasma
the brightness of some of the other sunlit scarps suggests that these reveal
cleaner or perhaps just more freshly exposed ice than is present across most
of the surface
the topographically subdued crater near the upper right, Gertrude, is 275 km
across and is the largest definite crater in the Uranus system

Oberon

The best image of Oberon is adequate only to show details
down to about 20 km in size (Figure 11.11). This is sufficient to
show that the whole of the visible area of its surface is about as
densely cratered as Umbriel. However, Oberon is less dark than

Umbriel, and there are patches of bright ejecta surrounding what are presumably the youngest impact craters. The darkest parts of its surface are patches on the floors of some of the craters.

figure 11.11   Voyager 2 image of Oberon (1520 km diameter)

Oberon has a notable mountain peak about 11 km high that is silhouetted against the blackness of space at the lower left of Figure 11.11. This is perhaps the central peak of a large impact structure whose rim has subsided to near invisibility because the lithosphere in the past could not support its weight. Detailed inspection of Oberon's profile suggests that the diameter of this subdued basin is about 375 km, and if this is real it would mean that Titania's crater Gertrude (Figure 11.10) is not the largest impact crater in the Uranus system.

One other noteworthy feature of Oberon that is visible in Figure 11.11 occurs near the day–night boundary at the upper right. This is a trough named Mommur chasma whose walls are scalloped rather than smooth in plan view, but which nonetheless could be a Titania-style fault trough. Once again we are left to speculate on whether the lack of likeness between two neighbours is more apparent than real. It is frustrating that we are likely to have to wait decades before we get images of the unseen portions of Uranus's fascinating satellites.

12 Neptune

In this chapter you will learn:
- about the outermost large planet in our Solar System, Neptune, and its strikingly blue and dynamic atmosphere
- about Neptune's slender rings and its family of satellites, one of which, Triton, is large and surprisingly active.

Planetary facts	
Equatorial radius (km)	24,766
Mass (relative to Earth)	17.14
Density (g/cm³)	1.64
Surface gravity (relative to Earth)	1.12
Rotation period	16.11 hours
Axial inclination	29.6°
Distance from Sun (AU)	30.11
Orbital period	163.7 years
Orbital eccentricity	0.009
Composition of surface	gassy
Mean cloud top temperature	–214 °C
Composition of atmosphere	hydrogen (79%), helium (18%), methane (3%)
Number of satellites	13

Neptune was the first planet to be discovered as the result of a deliberate hunt. At opposition Neptune's magnitude is 7.7 on the scale of stellar brightness used by astronomers. Magnitude 6 is about the limit for visibility with the unaided eye in a clear, dark sky, so Neptune cannot be seen without optical aid. Once sufficient time had elapsed since the chance discovery of Uranus for its orbit to be characterized, it became apparent that certain minor discrepancies in its path could best be explained by the presence of an eighth planet orbiting beyond Uranus. On this basis, the English mathematician John Couch Adams and the Frenchman Urbain LeVerrier each independently predicted the correct location of this planet. LeVerrier had better luck in persuading observers to look in the position he indicated and it was as a result of his work that Neptune was first seen by the Germans Johann Galle and Heinrich D'Arrest in 1846. Within a few weeks of its discovery, the Englishman William Lassell (who went on to discover Ariel and Umbriel in 1851) spotted Neptune's only large satellite, Triton.

Rotation and orbit

With an eccentricity of only 0.009 Neptune's orbit is more circular than that of any other planet except Venus. The Earth overtakes Neptune every 367.5 days, so that successive oppositions occur only about two days later each year.

Neptune's axial inclination of just below 30° is much more similar to Saturn's than to Uranus's. It rotates slightly faster than Uranus, but is slightly less flattened: its polar radius is only about 1.7 per cent less than its equatorial radius, implying a slightly more rigid interior than in the case of Uranus.

Neptune has 11 known satellites: six smallish inner ones travelling in virtually circular prograde orbits close to the planet's equatorial plane, then its only large satellite, Triton, in a retrograde orbit. Four outer irregular satellites are known, the largest of which, Nereid, has a prograde but highly elliptical orbit.

Missions to Neptune

Like Uranus, Neptune has been visited only by Voyager 2, and there are as yet no plans for a follow-up mission. Voyager 2 flew past Neptune and then Triton in August 1989. With no onward targetting requirement to constrain its path, the trajectory was chosen to optimize radio occultation studies of these two bodies, at the end of which the probe was travelling towards a direction 48° south of the ecliptic.

The interior

The structure of Neptune's interior is probably similar to that of Uranus. Neptune's slightly higher density implies a slightly less deep molecular hydrogen layer and a correspondingly larger core, where the pressure and temperature would nevertheless be similar to those within Uranus. Voyager 2 discovered that Neptune's magnetic field is about 25 times the Earth's, or half the strength of Uranus's, like which it is inclined at a large angle (47°) to the planet's rotational axis. Therefore, Neptune's field too is presumed to be generated by convection currents within a zone of liquid ice. In this case the field's centre of symmetry is offset even further from the planet's centre, just over halfway to the surface.

Neptune's cloud top temperature is the same as Uranus's, despite being half as far again from the Sun. This indicates that, in contrast to Uranus, Neptune is probably generating a significant amount of heat internally through contraction or differentiation.

The atmosphere

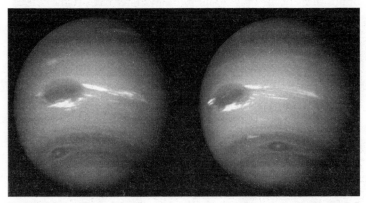

figure 12.1 two images of Neptune by Voyager 2, recorded about one planetary rotation apart in August 1989
the Great Dark Spot (22° south) has not quite completed a full rotation whereas the Small Dark Spot (54° south) has made a little over a full rotation of the planet
the Great Dark Spot was thus drifting westwards relative to the Small Dark Spot at about 100 metres per second

At the time of the Voyager fly-by, Neptune presented a considerably less bland appearance than Uranus, having clear latitudinal banding, two obvious dark spots in the southern hemisphere (rotating storm systems), and some high-altitude white clouds (Plate 11, Figure 12.1). Neptune's blue colour in contrast to Uranus's green appearance is probably because there is a greater depth of methane gas overlying Neptune's global methane cloud deck. Methane gas is mostly transparent at visible wavelengths, but it does absorb slightly in the longer-wavelength part of the visible spectrum. Therefore the opaque cloud deck below appears blue, just as the sandy floor of a clear tropical sea appears blue when seen from above the waves.

The Great Dark Spot looks superficially like Jupiter's Great Red Spot. Both are just over 20° south of their planet's equator and rotate anticlockwise, but there are some important differences. The Great Red Spot maintains a fairly constant outline, but Voyager 2 showed that the Great Dark Spot was changing shape and orientation with a period of about eight days. Furthermore, high-altitude methane cirrus clouds form above the Great Dark Spot as air rises and flows over and around the storm centre

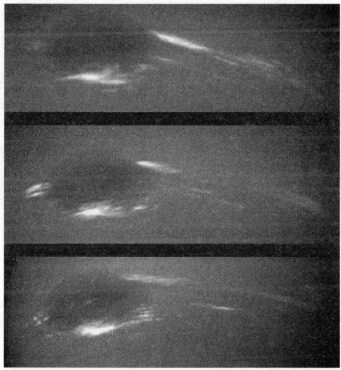

figure 12.2 Voyager 2 images of Neptune's 10,000-km-long Great Dark Spot recorded at intervals of 18 hours
the high bright clouds are analogous to terrestrial cirrus clouds, but are made of methane crystals rather than water-ice crystals

without becoming incorporated within it (Figure 12.2). There is no equivalent associated with Jupiter's Great Red Spot. At the time of the Voyager 2 encounter, the Great Dark Spot was drifting towards the equator at a rate of about 15° per year. We do not know if it ever reached it, but no surviving trace of the spot can be seen on Hubble Space Telescope images recorded in 1996 so it is unlikely that it was such a long-lived feature as Jupiter's Great Red Spot.

The Small Dark Spot seen on Neptune by Voyager 2 had more structure within and around it than the Great Dark Spot (Figure 12.3). Strangely, it appeared to be rotating clockwise, in the opposite direction to its larger counterpart.

figure 12.3 5000-km-wide Voyager 2 image of Neptune's Small Dark Spot, showing details as small as 20 km across
methane cirrus clouds are gathered over the spot's centre

Neptune's northern hemisphere was spot-free during the Voyager 2 fly-by, but shared the subtle banded appearance of the southern hemisphere accompanied by streaks of methane cirrus cloud (Figure 12.4). Hubble Space Telescope images obtained in 1998 showed even more prominent bright clouds present in this region.

figure 12.4 streaks of methane cirrus clouds up to 200 km wide and 29° north of Neptune's equator seen by Voyager 2 from a range of 157,000 km (two hours before closest approach to the planet)
shadows cast by these clouds onto the underlying cloud deck show that they were about 50 km above it

By tracking atmospheric features on Voyager 2 images it was possible to show that Neptune's atmosphere rotates in step with the planet's interior at about 50° north and south of the equator. At higher latitudes the atmosphere rotates as much as about 200 m per second faster than this, but near the equator it rotates over 300 m per second slower than the interior. Neptune thus shares with Uranus the property of westward-blowing equatorial winds (opposite to those on Jupiter and Saturn), despite the major difference in their axial inclinations.

It is surprising that Neptune's atmosphere is so dynamic. At Neptune, sunlight, and hence solar warming of the atmosphere, is 900 times less than at the Earth and over 30 times less than at Jupiter, so solar warming is likely to have little effect. Presumably atmospheric circulation is powered by heat escaping from the interior under the influence of the planet's rotation.

Rings

After the discovery of rings round Jupiter and Uranus, astronomers watched stellar occultations by Neptune very carefully to try to detect temporary dimming of the starlight that could be caused by passage of rings in front of the star. Several occultations passed without incident, but eventually a 10-km-wide ring was found during an occultation in July 1984. This marked the discovery of what we now know to be Neptune's outermost ring, but the occulation data showed it on one side of the planet only! The reason was revealed by Voyager 2, which showed that material in this ring, now named the Adams ring, is concentrated into discontinuous arcs (Figure 12.5). The reason why the ring appeared to be missing on one side of the planet when discovered by occultation, is that the star was crossed by one of the thinly populated 'breaks' in the ring, and so appeared undimmed.

Voyager 2 revealed a total of five rings around Neptune. Two of the others are narrow like the Adams ring and two are broad but diffuse. The innermost ring, the Galle ring, is one of the broad ones and has a radius of 41,900 km. The ring material is dark (albedo less than 0.1) and probably red, and most of it seems to be centimetres or metres in size.

figure 12.5 Voyager 2 image of part of Neptune's ring system, seen looking back towards the Sun
three clumps of material or 'ring arcs' are visible in the near part of the outer (Adams) ring at the top
inside this ring can be seen the LeVerrier ring
part of the deliberately overexposed crescent of Neptune is visible at the lower right

The clumping of the Adams ring into arcs was initially thought to be a gravitational effect resulting from the exact 42:43 orbital resonance between ring particles and the 160 km wide satellite Galatea, which orbits 980 km closer to Neptune, inclined at 0.03° to the ring plane. Telescope images of the Adams ring obtained for the first time in 1998 confirmed that the arcs were still present, but showed them in the wrong positions for resonance with Galatea to be the only controlling factor. Perhaps there is an undiscovered satellite (presumably less than 6 km across, or it would probably have showed up on Voyager images) orbiting within or close to the ring that helps to keep the arcs in shape.

Satellites

Neptune's known satellites are listed in Table 12.1. They have all been named after gods and nymphs connected with the sea. The six inner ones were revealed for the first time on Voyager 2 images in 1989, although it was subsequently realized that

Larissa had been detected in 1981 during a stellar occultation. The four innermost orbit among the ring system. These and the other two inner satellites appear to be covered by the same low albedo material as the rings. In this respect, they resemble the inner moonlets of Uranus except that Proteus and Larissa are larger than anything found there. Of the six, only the largest, Proteus, was seen close enough to reveal any details (Figure 12.6). Proteus is the largest planetary satellite whose shape is distinctly non-spheroidal, and is bigger than Mimas (Saturn) which is more regular in shape.

Name	Distance from planet's centre (km)	Radius (km)	Orbital period (days)	Mass	Density (g/cm³)
Naiad	48,227	29	0.294	?	?
Thalassa	50,075	40	0.312	?	?
Despina	52,526	75	0.335	?	?
Galatea	61,953	80	0.429	?	?
Larissa	73,548	104 × 89	0.555	?	?
Proteus	117,647	218 × 208 × 201	1.122	0.5	1.3
Triton	345,760	1353	5.877 R	2.14 × 10²⁰ kg	2.06
Nereid	5,513,400	170	360.14	0.2 × 10²⁰ kg	1.0
S/2002 N1	15,686,000	24	1875	?	?
S/2002 N2	22,452,000	24	2919	?	?
S/2002 N3	22,580,000	24	2982	?	?
S/2004 N4	46,570,000	30	8863R	?	?
S/2003 N1	46,738	14	9136R	?	?

table 12.1 satellites of Neptune (R = retrograde, ? = value unknown)

Unlike the other giant planets, Neptune has no family of large regular satellites orbiting beyond its inner moonlets. Instead there is just one large satellite, Triton, in a circular retrograde orbit inclined at 157°. Triton is discussed below, and is clearly a captured body whose arrival in a retrograde orbit would have scattered any satellites that were previously present in the conventional regular satellite zone around Neptune. These would have been smashed by collisions with each other or with Triton itself, or been ejected from orbit about Neptune. Neptune's original inner moonlets would probably have been involved in collisions at this time too, and it is possible that Proteus is a collisional fragment or a re-accreted body.

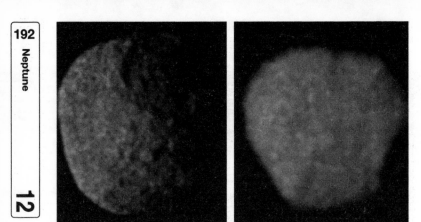

figure 12.6 two Voyager 2 views of Neptune's second-largest satellite, Proteus (420 km across)
both are good enough to reveal its non-spheroidal shape, and the one on the left, which is from closer range, shows more details and hints at a heavily cratered surface

Beyond Triton lies the only other satellite of Neptune known before the Voyager era. This is Nereid, which is smaller than Proteus but easier to see telescopically because it orbits much further from Neptune's glare and has a higher albedo, of about 0.2. Its orbit is inclined at 28° and has an eccentricity of 0.753, which is greater than that of any other known planetary satellite. Nereid is a body composed of dirty ice, and could be the remains of one of Neptune's original (pre-Triton) satellites, forced into an outlying, inclined and eccentric orbit.

Neptune has a family of small irregular outer satellites, like the other giant planets. They could either have been captured after Triton, or be survivors or products of the disruption that Triton's capture would have caused. Neptune's distance from the Earth makes such small objects even more challenging to detect than the ones in orbit around Uranus. At the time of writing only five have been found, all in eccentric and inclined orbits, two of which are retrograde (Table 12.1).

Triton

Triton was the last body visited by Voyager 2, which, hurtling past at a relative velocity of 27 km per second, imaged the sunlit part of its Neptune-facing hemisphere and revealed details as small as 400 m. Triton provided a fitting climax to the mission. Its basic properties are listed in Table 12.2. Triton has the coldest surface temperature (−235 °C) of any planetary body yet studied at close range, but despite this it shows evidence of a complex geological history, and has a variable atmosphere and a south polar cap punctured by geyser-like eruptions (Plate 12).

Equatorial radius (km)	1353
Mass (relative to Earth)	0.0036
Density (g/cm^3)	2.05
Surface gravity (relative to Earth)	0.080
Composition of surface	ice (nitrogen, carbon dioxide, methane, carbon monoxide)
Mean surface temperature	−235 °C
Composition of atmosphere	nitrogen (>99%), methane, carbon monoxide
Atmospheric pressure at surface (relative to Earth)	1.4×10^{-5} (1989) 3×10^{-5} (1997)

table 12.2 planetary facts for Triton

Given that Triton is a captured body, its only likely provenance is from the Kuiper belt. Thus, study of Triton can give us insights into the composition of these as yet unvisited bodies. However, Triton is exceptional because it is larger than any known object still belonging to the Kuiper belt, and because its capture by Neptune may have involved it in violent collisions with other satellites. Capture must have been followed by a prolonged period of tidal heating, perhaps lasting a billion years before Triton's orbit became circular and its rotation synchronous with its orbital period.

Triton's density is higher than that of any other outer planet satellite except Io and Europa, and indicates that rock (plus iron if this has separated out into a differentiated inner core) makes up about two-thirds of its total mass. Triton's average albedo is high, about 0.8, indicating a bright icy surface. Spectroscopic studies by telescope show that the surface is dominated by nitrogen ice, but ices formed from methane, carbon dioxide, carbon monoxide and water have been detected too. It is difficult to determine relative proportions of the compounds in the surface ice, because nitrogen and possibly methane and carbon monoxide also occur in the atmosphere. Furthermore, their abundances and distributions vary seasonally.

Triton's seasons are peculiar, because they result from a combination of Neptune's 29.6° axial inclination and Triton's

21° orbital inclination. Triton's orbital plane precesses about Neptune's axis, so the full seasonal cycle on Triton equates not to Neptune's 164-year orbital period but to a 688-year cycle, with 164-year subcycles superimposed. During the 688-year cycle the subsolar latitude on Triton varies between extremes of 50° N and 50° S. As chance would have it, at the time of the Voyager 2 fly-by, Triton was approaching its extreme southern summer with the Sun overhead at nearly 50° S, so a large part of Triton's northern hemisphere was in long-term darkness and could not be seen. At this time Triton's south polar cap of bright nitrogen ice (albedo 0.9) was in retreat, subliming directly to nitrogen gas. This probably explains why the atmospheric pressure doubled from 14 millionths of the Earth's atmospheric pressure in 1989, as determined by Voyager 2 radio occultation, to about 30 millionths when estimated telescopically by stellar occultation in 1997.

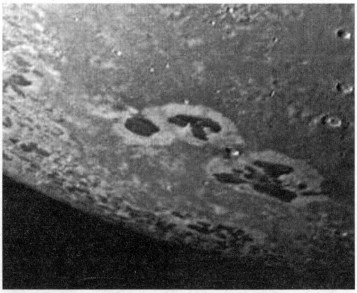

figure 12.7   Voyager 2 view of a 600-km-wide region on Triton, which lies just out of sight below the bottom of Plate 12
the ragged edge of the retreating polar cap of nitrogen ice is visible on the left
the pale zones around the dark patches near the centre could be remnant polar ice, or could be something that condensed out around cryovolcanic flows represented by the dark patches
only a few impact craters are visible, of which the largest (on the right) is 27 km across
this is part of Triton's leading hemisphere, and so is more vulnerable to impacts than the trailing hemisphere

Apart from the morphological evidence for seasonal retreat of the south polar cap (Figure 12.7 and Plate 12), four jets of gas were actually seen escaping from the cap. These reached about 8 km in height, and were made visible because of dark particles suspended within them. What appears to have been happening is that sunlight was penetrating the thin cap of pure nitrogen ice (which would have condensed out of the atmosphere during the previous winter) and was being absorbed by the darker substrate. Thus the substrate was warmed by the Sun. Heat conducted upwards from the substrate into the base of the polar cap caused this to sublime directly to gas, which then escaped under pressure through fissures or vents in the ice, carrying with it dusty dark stuff (possibly carbon or tholins) from the substrate. On reaching a height where they are no longer buoyant enough to rise, these geyser-like jets become deflected sideways by the prevailing high-altitude winds, estimated to be blowing at about 15 m per second. Jets are too small to make out in Plate 12, but plenty of dark streaks can be seen on top of the polar cap. These correspond to the down-wind plumes themselves or else to the stains on the ice left by particle fall-out from the plumes.

Other notable features of Triton's atmosphere include rare thin clouds at an altitude of a few km that are thought to consist of

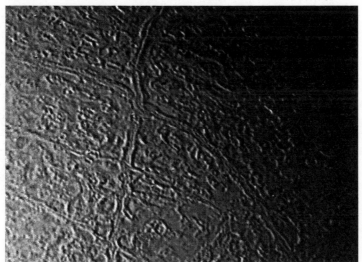

figure 12.8 Voyager 2 view of a 600-km-wide part of Triton's cantaloupe terrain lying within the upper right of Plate 12
the intersecting ridges, forming an X-pattern, are among the youngest features on Triton, and are probably sites where viscous cryovolcanic magma has welled up along fissures

condensed particles of nitrogen-ice, making them nitrogen versions of Earth's cirrus clouds. A tenuous haze extends above this level to a height of 30 km. This may be a photochemical smog containing such molecules as ethane, acetylene and hydrogen cyanide created by the action of solar ultraviolet radiation on methane and ammonia, like the much thicker smog on Titan.

Triton's surface that is visible where the polar cap is absent presents a remarkably varied appearance. There are few impact craters, attesting to the young age of the surface. Most of the craters probably reflect impacts by small (100–700 metre) bodies from the Kuiper belt. Calculations based on the expected rate of such collisions suggest that the average age of Triton's surface is no more than a surprisingly young 300 million years. Less cratered parts of the surface may be only tens of millions of years old. If true this makes Triton the most active known solid body in the outer Solar System except for Io and probably Europa. One school of thought attributes this to residual tidal heating reflecting relatively recent capture of Triton by Neptune, perhaps within the past billion years or so. Another suggests that we are seeing the effects of the ability of radiogenic heating to power low-temperature cryovolcanism in Triton's volatile cocktail of ices.

A particularly weird region on Triton occupies much of the upper right in Plate 12, and part of this is shown in more detail in Figure 12.8. Because its texture resembles the skin of a cantaloupe melon, this has been dubbed the cantaloupe terrain. Here, the surface is pockmarked by dimples about 30 km across. In their narrow range of sizes and detailed morphology these are quite unlike impact craters. They may represent places where pods of warm slushy buoyant ice rose towards (and maybe breached) the surface. This would be analogous to diapirs on Earth of the kind that occur when 10-km-sized pods of salt rise towards the surface, piercing through the overlying sedimentary rocks.

Near the boundary between leading and trailing hemispheres the cantaloupe terrain is overlain by smoother plains units. These are almost certainly products of cryovolcanic flooding. Given that nitrogen, methane, carbon dioxide, carbon monoxide and water have all been detected and that ammonia is probably present too, the range of possible magma compositions is at least as great as on Earth. Nitrogen would vaporize at a relatively low temperature, and its abundance on

Triton makes it likely that some eruptions would have happened in an explosive fashion, driven by the violent escape of nitrogen gas. In parts of this region there are smooth-floored and irregular-walled depressions, that may indicate that the previous surface collapsed or was melted (Figure 12.9).

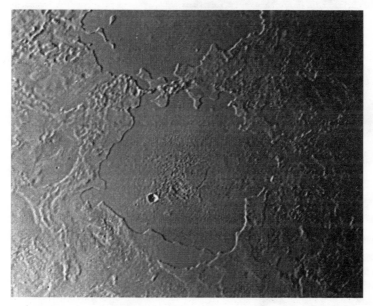

figure 12.9 Voyager 2 view of a 300-km-wide region of Triton's cryovolcanic plains the 180-km-wide depression in the middle, set into the surrounding high plains, is named Ruach Planitia
its walls are about 200 m high

The youngest features of Triton's permanent surface are complex linear ridges, such as those seen crossing the cantaloupe terrain in Figure 12.8, some of which can be traced for over 1000 km. These may simply be sites where viscous cryovolcanic magmas have oozed out of fissures, or they could have more complex origins such as those suggested for the ridges on Europa.

It would be fascinating to return to Triton, to discover how much of the south polar cap has vanished and to find out what lies in the previously unseen parts of the globe. However, the next mission to the fringe of the Solar System is likely to be sent to Pluto. This is probably an equally exciting place, as discussed in the next chapter.

13

Pluto, Charon and the Kuiper belt

In this chapter you will learn:

- about the planetary bodies on the outer fringes of our Solar System
- that Pluto although still officially classified as the outermost 'planet' is really just the largest known object in an extensive zone of icy bodies known as the Kuiper belt
- that Pluto has a large satellite, a property now known to be shared with several other Kuiper belt objects.

Planetary facts for Pluto and Charon		
	Pluto	**Charon**
Equatorial radius (km)	1150	625
Mass (relative to Earth)	0.0022	0.0003
Density (g/cm³)	2.0	1.7
Surface gravity (relative to Earth)	0.059	0.021
Rotation period	6.39 days	6.39 days
Axial inclination	119.6°	0° (relative to orbit about Pluto)
Distance from Sun (AU)	39.5	39.5
Orbital period	248 years	6.39 days (about Pluto)
Orbital eccentricity	0.249	0.002
Composition of surface	icy (nitrogen, methane, carbon monoxide, ethane)	icy (mostly water)
Mean surface temperature	−230 °C	−230 °C
Composition of atmosphere	mostly nitrogen, possibly also methane, carbon monoxide and ethane	none known
Atmospheric pressure at surface (relative to Earth)	5×10^{-5}	
Number of satellites	1	0

Like Neptune, Pluto was discovered as the result of a deliberate search. Towards the close of the nineteenth century, apparent minor deviations in the orbits of both Uranus and Neptune pointed to the existence of an unknown planet orbiting beyond the then known Solar System. The American astronomer Percival Lowell (and others) made repeated attempts to locate this hypothetical planet, which became known as Planet X, in the early years of the twentieth century. Lowell died in 1916, but in 1929 the young American Clyde Tombaugh was recruited to continue the search, at the Lowell Observatory in Flagstaff, Arizona. Tombaugh compared vast numbers of star-like objects on photographic plates covering the appropriate part of the sky, which had been taken a few nights apart in order to reveal the planet by its motion relative to the background stars. In 1930, on plates taken on 23 and 29 January of that year, he found a faint (magnitude 14) object within a few degrees of the predicted position and which had moved by the amount expected of a trans-Neptunian planet. The newly discovered planet was aptly named Pluto after the classical god of the underworld, but it is fitting that the first two letters of its name are Percival Lowell's initials.

Pluto is too far and too faint for much to be learned about it for most of the next half-century, even its size being a mystery. Prior to discovery Pluto was assumed to be about ten Earth-masses, and upon discovery its faintness caused its estimated mass to drop tenfold. However, after Pluto's satellite, Charon, had been discovered in 1978, the characterization of Charon's orbit proved that Pluto's mass is only about a quarter of a per cent of the Earth's. This revealed it as an icy body of far too little mass to be responsible for the apparent deviations in the orbits of Uranus and Neptune. We now know that these deviations were not real, but instead reflect tiny errors in nineteenth-century observations of planetary positions.

Pluto's discovery near the expected position of a massive trans-Neptunian planet, although a just reward for painstaking and dedicated work, was thus essentially a matter of luck. The post-1992 documentation of hordes of Kuiper belt objects orbiting between 30 and 50 AU means that it is quite impossible for there to be an undiscovered major planet in the same region of space.

The realization that Pluto is much smaller than originally assumed and shares many properties with Kuiper belt objects led to a lot of media speculation in 1998–1999 that Pluto was about to be stripped of its planet status, and become officially classified as just another Kuiper belt object. However, the International Astronomical Union acted to quash this rumour, and on the basis of historical precedent Pluto will continue to be listed as a planet for the foreseeable future.

Rotation and orbit

During the period 5 September 1979 to 11 February 1999 it was a trick question to ask 'What is the furthest planet from the Sun?' because, although Pluto's average distance from the Sun is greater than Neptune's, between those dates the correct answer was Neptune. This is because Pluto has a more eccentric orbit than any other planet, and when it is near perihelion it is closer to the Sun than Neptune. Pluto was only at 29.66 AU from the Sun at its most recent perihelion on 5 September 1989. Its distance will now increase until it reaches aphelion at 49.54 AU in 2114.

Pluto's orbital period is exactly 50 per cent longer than Neptune's, which means that the two are in 3:2 orbital resonance. Although Pluto's orbit comes inside Neptune's so that their orbits cross when drawn in plan view (e.g. Figure 2.2), Pluto's orbit is inclined to the ecliptic by 17° and their paths do not actually intersect. This means there is no possibility of a mutual collision. Pluto's inclined orbit takes it 8 AU north of the ecliptic at perihelion, and 13 AU below it at aphelion, and its orbital resonance with Neptune results in the distance between the two bodies being never less than 17 AU.

Like Uranus, Pluto is a planet that lies on its side, having an axial inclination of nearly 120°, and therefore retrograde rotation. This did not become fully apparent until after the discovery of Charon, which orbits in Pluto's equatorial plane with a period of 6.4 Earth-days. Imaging by modern telescopes has revealed that not only does Charon keep the same face to Pluto (being in synchronous rotation, like most planetary satellites), but Pluto itself always keeps the same face towards Charon. The two bodies are mutually tidally locked, so that Pluto's rotation keeps pace with Charon's orbital motion about it. However, Charon's orbit is not exactly circular, having an eccentricity of about 0.002 (possibly a result of a relatively recent collision between Charon and a body of comet size or larger), so there must be some tidal heating going on today.

Pluto and Charon are probably a product of a giant impact between two Kuiper belt objects, in the same way that the Earth-Moon pair results from a giant impact between planetary embryos, except that the collision knocked Pluto onto its side. Pluto and Charon are the nearest thing the Solar System can boast to a double planet, an honour formerly thought to belong to the Earth and Moon. Charon's 19,400 km orbital radius is only about 16 times the radius of its planet, whereas the Moon's orbital radius is about 60 times the Earth's radius. Moreover, Charon has 12 per cent the mass of Pluto whereas the Moon has only 1.2 per cent the mass of the Earth. Strictly speaking, no satellite orbits around the centre of its planet, but about the system's centre of mass (or 'barycentre'). Usually, this point lies inside the planet, but Charon's mass is such a large fraction of Pluto's that their barycentre lies between the two bodies in open space. It does, however, lie closer to Pluto than to Charon, in proportion to Pluto's greater mass. Thus, it is more realistic to regard both bodies as orbiting their barycentre, Pluto with a smaller orbit and Charon with a larger orbit such that the average Pluto–Charon distance corresponds to the orbital radius quoted above.

The best images of this remarkable pair are shown in Figure 13.1. Repeated searches for smaller satellites of Pluto have failed to find any, and if Pluto does have any other companions, none can be larger than about 100 km across.

Missions to Pluto and beyond

Pluto is the only planet not yet visited by a spaceprobe. Various NASA missions to Pluto have been proposed and then dropped through lack of funding. The current plan is for a mission called New Horizons to be launched in January 2006. This could reach Pluto as soon as November 2015, after a gravity assist fly by of Jupiter. The probe would collect visible, infrared and ultraviolet images and spectroscopic data with the aim of determining surface and atmospheric compositions of Pluto and Charon. It would pass within 27,000 km of Charon and within 10,000 km of Pluto. The best images would show details less than 100 m across, but most images would be at considerably poorer resolution. If the spacecraft remains healthy, it will be directed onward for an encounter with one or more Kuiper belt objects.

Pluto's interior

We have no direct information on the interiors of Pluto or Charon. However, because of the melting that would be caused by a giant impact and the heating that must have occurred while tidal forces were bringing Charon into a near-circular synchronous orbit, it is very likely that both have become internally differentiated. Pluto's density suggests that below its icy mantle there is a rocky core containing about 70 per cent of the planet's mass, with possibly an iron-rich inner core.

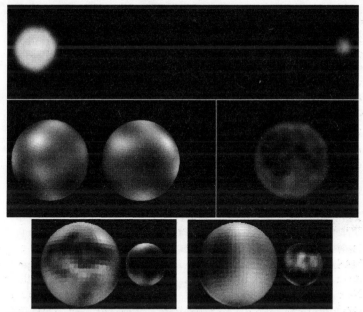

figure 13.1   a collage of the best images of Pluto and Charon
top row: Hubble Space Telescope image recorded in February 1994 showing
their true relative sizes and separation
middle row: views of Pluto's western and eastern hemispheres constructed by
addition and enhancement of Hubble Space Telescope images obtained in
June and July 1994 (left), and, for comparison, an image of the near side of
the Moon degraded to similar resolution (right)
bottom row: albedo patterns on Pluto and Charon, left Charon-facing side of
Pluto and anti-Pluto side of Charon, right anti-Charon side of Pluto and Pluto-
facing side of Charon
these maps were constructed on the basis of variations in the combined
brightness of the two bodies as they rotate, in particular during mutual
occultations in 1985–1990 when the orientation of Charon's orbit with respect
to Earth was such that it repeatedly passed in front of and behind Pluto

Pluto's atmosphere and surface

It is thanks mainly to modern telescopic and image-processing
techniques that we have any knowledge at all of Pluto's surface
markings (Figure 13.1). Although Neptune's largest satellite,
Triton, is the most similar planetary body we know (and

appears to be an escapee from the Kuiper belt, and thus Pluto's larger cousin), there are clearly some important differences. Pluto's surface is bright but with a surprising degree of contrast between its brightest regions, where the albedo is as high as 0.7, and the darkest spots, where albedo drops to about 0.15. Spectroscopic studies have revealed that the surface consists of frozen nitrogen (presumably concentrated as frost in the brightest regions), methane and carbon monoxide. There are also traces of ethane apparently dissolved within the nitrogen ice. The dark areas are probably tholin-rich residues. Because they absorb more solar heat than the bright areas, they have a temperature about 20 degrees higher than these, and are warmer than anywhere on Triton (at least while Pluto is near perihelion). The bulk of Pluto's ice is likely to be water in composition, but this has not been detected spectroscopically. Presumably it lies buried beneath the more volatile ices even more completely than is the case on Triton.

Pluto's atmosphere was confirmed when the planet passed in front of a star in 1988. It is assumed that nitrogen, the most volatile of the surface ices, makes up the bulk of the atmosphere, but that methane, carbon monoxide and ethane are present too. The temperature contrast between the dark and bright regions is likely to drive ferocious near-surface winds. There is perhaps a tenuous low-altitude haze layer caused by such substances as hydrogen cyanide, acetylene and ethane, that could be a photochemically induced smog like those on Titan and Triton.

Although Pluto's atmosphere is insubstantial, the planet's feeble gravitational hold means that it is particularly extensive. For example, an imaginary shell enclosing 99 per cent of Pluto's atmosphere would be about 300 km above the surface, whereas for the Earth the equivalent height is only 40 km. The outermost traces of Pluto's atmosphere probably occupy a sphere similar in radius to the Earth itself.

Near the time of perihelion, Pluto's atmospheric pressure was about 5 millionths of the Earth's, which is comparable to that of Triton. However, because of the eccentricity of Pluto's orbit, the planet receives nearly three times more solar warmth at perihelion than it does at aphelion. Added to the seasonal changes resulting from the planet's axial inclination, this means that Pluto's mean temperature must be even lower than its perihelion value for much of its orbit. As it moves further from the Sun (and, coincidentally, one pole moves into season-long shadow) much of Pluto's present atmosphere is vulnerable to becoming frozen out onto the surface. If this happens, the atmospheric pressure will plummet as Pluto approaches aphelion. On the other hand the greenhouse properties of the atmosphere could be sufficient to keep the globe warm enough to avoid this fate.

So, what will the first close-up images of Pluto show us? If we have learned anything from exploring the Solar System, it should be to expect the unexpected. However, the most likely situation is that Pluto's surface features will turn out to have many characteristics in common with those of Triton. Both bodies appear to have originated in the Kuiper belt and have similar substances on their surfaces, both were probably involved in major collisions, and both must have experienced tidal heating over a prolonged period. Thus Pluto probably has a diversity of terrain types, including many cryovolcanic landforms. Substantial tracts can be expected to be partly obscured by nitrogen frost deposits, and this frost cover will probably increase in extent as Pluto moves away from perihelion.

Charon

Charon is named after the ferryman who, in classical mythology, transported the ghosts of the dead across the river Styx into the underworld domain of the god Pluto. It has a patchy surface, like Pluto (Figure 13.1), but it is darker and the only ice species to have been detected is water. The dark contaminant in its ice is neutral grey rather than red, and so it is probably rock or soot, rather than tholins.

Charon appears to have no atmosphere. Perhaps Charon was once equally as rich as Pluto in ices of nitrogen, methane and carbon monoxide. However, Charon's gravity is too weak to hold onto gas molecules, and so some would have escaped to space during each perihelion passage when the temporary relative warmth would have turned some of the ice to gas. Another factor that could explain their absence is that such volatile species would have more easily escaped Charon when it was accreting, especially if a giant impact event was involved in Charon's origin.

Thus while the mobility of Pluto's surface ices may make its surface relatively young, Charon's less volatile water-ice surface may turn out to be much older and heavily cratered. However, it has been suggested that the high inclination of Charon's orbit relative to the Sun could lead to competing tidal pulls on Charon by Pluto and the Sun of sufficient magnitude to cause melting within Charon's icy layer. If this is the case, then the intriguing prospect opens of a Europa-like surface for Charon with even a possibly life-bearing ocean beneath it. Let's hope we don't have to wait too long to find out.

The Kuiper belt

The existence of the Kuiper belt was first postulated on theoretical grounds by Kenneth Edgeworth in 1943 and (apparently independently) by Gerard Kuiper in 1951. It is thus sometimes referred to by the longer name of the Edgeworth-Kuiper belt, and sometimes the words 'disk' or 'cloud' are substituted for 'belt'. The term 'Kuiper belt object' is sometimes abbreviated to KBO, and its alternative, 'Edgeworth-Kuiper object', to EKO. This inconsistency is because the terminology is still evolving. Presumably in time scientists will conform to a standard vocabulary. It is to be hoped for the sake of everyone's sanity that the acronym EKO, pronounced as in the English word 'echo', does not take root, because it invites such appalling puns as 'distant EKOs', 'faint EKOs' and (for those with satellites) 'double EKO'.

Until the first Kuiper belt object (other than Pluto-Charon) was discovered in 1992, few people had been aware of the belt's hypothetical existence. It was decided that newly-discovered Kuiper belt objects should use the same provisional designation coding as asteroids, and so the first discovery happened to be designated 1992 QB_1. Five more were found in 1993, 12 in 1994, and so on at an accelerating rate as telescope time was allocated to the hunt. The overall tally is likely to pass one thousand overall before the end of 2004. At the time of writing few have been given names, and the majority are still identified by their provisional designations.

Dynamics of the Kuiper belt

As noted in Chapter 02, there are probably about 70,000 Kuiper belt objects more than 100 km in size. The acceptance that the Kuiper belt is real and contains a few tenths of an Earth-mass of primitive, thermally unprocessed, icy objects that escaped involvement in planetary formation has revolutionized our understanding of the history of the outer Solar System. Probably, the belt originally contained more than ten Earth-masses of material, but the gravitational influence of Neptune scattered most of these objects either inwards (those that escaped collision becoming Centaur asteroids or short-period comets) or outwards (mostly to be lost or to become long-period comets).

What remains of the Kuiper belt today is quite well structured. About a sixth of known Kuiper belt objects are in eccentric orbits in 3:2 resonance with Neptune. They share this characteristic with Pluto, and in consequence are sometimes referred to as Plutinos. A much smaller number of Kuiper belt objects have other orbital resonances with Neptune, including 2:1, 4:3 and 5:3. About half the known Kuiper belt objects have non-resonant orbits, with eccentricities of less than 0.15 and average distances from the Sun of between 41 and 49 AU. 1992 QB_1 was the first of this class to be identified, and so these objects are sometimes described as Cubewanos (QB_1-ohs) or less imaginatively as 'Classical Kuiper Belt Objects'. Most of the rest have inclined and eccentric orbits taking them well beyond the classical Kuiper belt at aphelion but in close to Neptune's orbit at perihelion. These belong to the 'scattered disk', which may exceed the classical Kuiper belt in total mass, and they presumably began life closer to Neptune but were flung outwards after a close encounter. Objects here are referred to as 'Scattered Disk Objects', and the term '**Trans-Neptunian object**' or '**TNO**' is used as a catch-all term to include these and true Kuiper belt objects.

Sizes and compositions

Because of their distance, it is exceptionally difficult to study Trans-Neptunian objects, which are typically fainter than 20th magnitude. Estimates of their sizes depend on assuming a value for their albedo, but some are now known that rival Charon in size (Table 13.1).

Spectroscopic studies have so far been adequate to indicate the presence of frozen water and methane on the brightest Kuiper belt objects, and to show that in general Kuiper belt objects are either neutral in colour or very red. This marked dichotomy in colour type is independent of the body's size or nature of its orbit. The red objects are probably coloured by tholins, whereas the grey ones are probably ice darkened by silicate dust or soot.

Object	Radius (km)	Distance from Sun (AU)	Orbital period (years)	Orbital eccentricity	Orbital inclination
Sedna	750	532	12,270	0.857	11.9°
2004 DW	750	39.5	248	0.218	20.6°
Quaoar	630	43.3	285	0.036	8.0°
Ixion	600	39.3	247	0.243	19.7°
2002 AW$_{197}$	600	47.5	328	0.128	24.3°
Varuna	450	43.3	285	0.053	17.1°

table 13.1 the largest Trans-Neptunian objects (other than Pluto and Charon) that were known in 2004
Ixion and 2004 DW are Plutinos, Quaoar and Varuna are Cubewanos, and 2002 AW$_{197}$ is a Scattered Disk Object.
Sedna (2003 VB$_{12}$) may be the first-known example of a new class of bodies, and is only briefly within the main part of the Kuiper belt when near its 44 AU perihelion.
Most of the sizes quoted are estimates assuming surface albedo of about 0.1.

Trans-Neptunian objects with satellites

As recently as 2001 close inspection of telescopic images of the Cubewano 1998 WW$_{31}$ revealed it to be a double object, and by the end of 2002 seven more had been identified (Table 13.2). These are challenging observations to make with ground-based telescopes, but a much clearer view can be obtained from space (Figure 13.2).

So far, none of the known satellites is very much smaller than the body it orbits. The satellite of 1998 WW$_{31}$ has about 85 per cent as much mass as its primary, and 2001 QW$_{322}$ is a pair of twins with no perceptible size difference between primary and satellite. It is not really surprising that the pairs with larger satellites should be identified first, because smaller satellites are harder to detect.

Eventually, study of the satellites' orbits should greatly improve our understanding of the masses, densities and sizes of Trans-Neptunian objects. At present about 1 per cent of the population is known to have satellites, but the proportion seems likely to rise as researchers intensify their searches. The existence of so many satellites is challenging to understand, and doubts have been raised over the likelihood of the majority having been created by giant impacts as proposed for Pluto-Charon.

In addition to having a large satellite, the Scattered Disk Object 1998 SM_{165} is noteworthy because its brightness doubles every four hours and then drops back to its original value. This is surely related to its rotation period. Either one side is twice as bright as the other, or else it is an elongated object rotating with a period of eight hours. The latter is more likely, and is reflected by the size estimate in Table 13.2. If correct, this is the largest irregular shaped icy body known, and perhaps indicates that 1998 SM_{165} has been shattered by a relatively recent collision.

Object	Radius of main object (km)	Radius of satellite (km)	(km)	Separation period	(days)	Orbital
Pluto-Charon	1150	625		19,660	6.39	
2001 QC_{298}	200	?		5000 ?	?	
1998 SM_{165}	300 x 150	100		6000?	15?	
1997 CQ_{29}	150	?		8000?	50?	
1998 WW_{31}	75	60		22,300	574	
1999 TC_{36}	350	50		8000?	11?	
2000 CF_{105}	100	70		23,000?	400?	
2001 QT_{297}	200	150		25,000?	150?	
2001 QW_{322}	70	70		120,000?	6000?	

table 13.2 Trans-Neptunian objects with known satellites apart from Pluto-Charon, these are Cubewanos, except for 1998 SM_{165}, which is a Scattered Disk Object, and 1999 TC_{36} and 2001 QW_{322}, which are Plutinos

Beyond the Kuiper belt?

There was great excitement in the spring of 2004 when the orbit was announced of a large TNO provisionally named 2003 VB_{12}, because although its perihelion lies within the classical Kuiper belt its aphelion is at 990 AU. This body was unofficially dubbed 'Sedna' (after an Inuit sea goddess), and for a while the world's media speculated that this is the Solar System's tenth planet. However, it is probably no bigger than the largest less-exotic Trans-Neptunian objects (table 13.1). It remains to be seen whether there are other Sedna-like object constituting an extended scattered disk.

Unless current theories of planetary origin are badly flawed (which is not impossible) the scattered disk should mark the limit for any planet-sized body orbiting the Sun and originating within the Solar System. However, in 1999 the British planetary scientist John Murray noticed an alignment in the aphelion positions of long-period comets and suggested that they could have been dislodged from their places in the Oort cloud by the passage of a planet-sized object (perhaps of Jupiter-mass or greater). He calculated that the best solution from the available data gave a retrograde orbit for this hypothetical planet, with an inclination of 150° and a mean distance from the Sun of about 32,000 AU, giving it an orbital period of nearly 6 million years. To detect this new Planet X by telescope would be hard but not impossible.

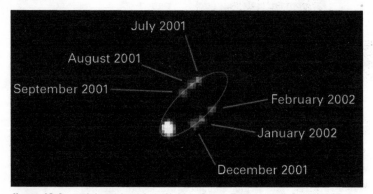

figure 13.2 a time series of six Hubble Space Telescope views of 1998 WW$_{31}$ made by superimposing successive images of the main body, to reveal the relative orbital motion of its satellite
orbital eccentricity is about 0.8
in reality, both objects orbit their common centre of mass (barycentre)
the two bodies are much more similar in size than implied by their relative brightnesses on this image, because the brightness of the main body has been exaggerated sixfold by the summing

The Sun's gravity is so weak at this extreme range suggested for Planet X that it is unlikely to have been able to retain such a planet since the origin of the Solar System. Thus, if it exists, this planet is more likely to be something captured by the Sun during the past billion years or so, rather than a long-term and permanent member of the Sun's family. At present, we have no knowledge at all of such interstellar planetary wanderers. However, planets are now known around several other stars, and these are the subject of the final chapter.

14

planets around other stars

In this chapter you will learn:

- that planets are common around Sun-like stars, but that the examples so far discovered are rather unlike the Solar System
- that this could be simply because our current method of detecting extrasolar planets works best for massive planets in close orbits
- ways to detect life on some of these planets.

It is now clear that planetary systems are common around stars other than the Sun, although it is too early to tell whether our own Solar System is typical. This is a major advance on the state of our knowledge only a few years ago.

Evidence that most young Sun-like stars have a surrounding ring of dust began to accumulate from the late 1970s onwards. Initially the evidence was merely the influence of the dust on the star's infrared spectrum, but by the mid-1980s it became possible to image the dust discs themselves. These discs, it was argued, could be direct analogues of the solar nebula prior to planetary formation (Figure 2.3). Alternatively, they could be dust surviving in the equivalent of each star's Kuiper belt (and beyond) after planet formation was complete in the inner part of each system. The first planet of another star (otherwise known as an extrasolar planet or **exoplanet**) was discovered as recently as 1995. The number of known exoplanets reached 100 in September 2002, and the total was rising rapidly.

None of these exoplanets has yet been seen directly. Most have been detected by the slight oscillation of the star's position as star and planet move around their common centre of mass (barycentre) in response to the planet's orbital motion. The star's motion is generally too slight to be visible as an actual change in its position. Rather the motion is inferred from inspection of the lines in the star's spectrum, whose wavelength depends on the star's velocity relative to the Earth. This is the phenomenon known as the 'Doppler shift', and is analogous to the changing pitch of a car's horn as it passes by. Wavelengths become shorter (higher pitched) if the source is approaching, and longer (deeper pitched) if the source is receding. Before this effect can be used to detect the star's wobble about the star-planet centre of mass, the additional, and continually changing, Doppler shift in the star's spectrum caused by Earth's own motion relative to the star has to be eliminated. Also the characteristics of the spectrometers used must be extremely stable. Thus the procedure is not straightforward, and requires very precise measurements. Furthermore, it is best at detecting really massive planets orbiting very close to their stars. The lower a planet's mass, or the further it is from its star, the harder it is to detect.

What this method tells us is how the star's velocity towards or away from us (known as radial velocity) varies in response to its planet's orbital motion. It does not detect any displacement of the star side-to-side of our line of sight. Thus it is described as the radial velocity method of detecting exoplanets.

Unfortunately, to determine the planetary mass we usually have to assume that plane of its orbit lies exactly in our line of sight, which in reality is rarely likely to be the case. Thus the planetary mass deduced by the radial velocity method is usually an underestimate, though typically by less than 50 per cent, except in the rare cases where the inclination of the orbit is known by other means.

Most known exoplanets have masses greater than Jupiter's. By convention, only those of less than 13 Jupiter masses are classed as planets. The centre of any object more massive than this would probably be at sufficiently high temperature and pressure for nuclear fusion of deuterium ('heavy' hydrogen) to begin. This would qualify the object as a partially failed star known to astronomers as a brown dwarf, and it would shine like a dim star for a few hundreds of millions of years before all its deuterium was used up. In order for nuclear fusion of ordinary hydrogen to begin, which is required to make a real star, a body's mass needs to be about 80 times that of Jupiter (or about 8 per cent the mass of the Sun).

At the time of writing, the least massive planet revealed by the radial velocity method orbits a Sun-like star known as HD 49674. If the orbit is more-or-less edge on to our line of sight, the planet has about 12 per cent the mass of Jupiter. This is less than Saturn but twice the mass of Uranus and Neptune (and more than that if the orbit is not nearly edge-on to us). Fairly clearly this is a giant planet, but it orbits its star at a distance of only 0.057 AU, which is far closer than Mercury is to the Sun. Such a close orbit is not untypical of those discovered by the radial velocity method, which is particularly good at detecting massive planets in close orbits because this is the situation needed to impart the greatest wobble to the star.

The radial velocity method has so far revealed a few stars with two planets, and two stars with three planets (Table 14.1). These systems look rather unlike our own Solar System. Clearly a giant planet very close to its star must be far hotter than those with which we are more familiar. It is thought unlikely that these could have formed so close to their star. Instead they probably grew sufficiently far out for it to be cold enough for ice to condense, and migrated inwards later. Recognition of this as a widespread process may cause us to reconsider the current model for how our own Solar System evolved. However, we need to develop ways to detect less massive exoplanets and those in more distant orbits, which would represent systems more like our own, before we can judge whether or not our Solar System is unusual.

Star	Planet	Mass (relative to Jupiter)	Average distance from star	Orbital period (days)	Orbital eccentricity
Upsilon Andromedae	b	0.71	0.059	4.617	0.034
	c	2.11	0.83	241.2	0.18
	d	4.61	2.50	1267	0.41
55 Cancri	b	0.84	0.11	14.65	0.02
	c	0.21?	0.24?	44.28?	0.34
	d	4.04	5.9	5360	0.16

table 14.1 exoplanetary systems with more than two planets
masses are quoted assuming the orbital plane is edge-on to our line of sight, and are likely to be underestimates
by convention, the planets are labelled b, c and d, in order of discovery

One such way to detect exoplanets is to note the slight dimming of a star's brightness as one of its planets passes in front of it during an occultation. This requires that the plane of the exoplanet's orbit is almost exactly edge-on as seen from Earth, the chances of which are about one in 200. Furthermore, most planets are so small that they can hide only a tiny fraction of their star's disc, so the dimming effect is subtle and hard to measure through the Earth's atmosphere. By 2002, only one planet had been discovered in this way; a body with 0.7 times the mass of Jupiter but 1.5 times Jupiter's radius, orbiting the star HD 209458 at 0.045 AU with a period of 3.5 days. Even this relatively large planet blots out less than 2 per cent of its star's light when it passes in front of it. To detect smaller and potentially more Earth-like planets in this way probably requires a telescope in space. There is a proposal for a NASA space-mission called Kepler to deploy a one metre aperture telescope for this purpose, perhaps as early as 2006. By continuously monitoring the brightness of 100,000 Sun-like stars over four years, Kepler would be expected to find about 500 Earth-like planets, if these are at all common.

To record an actual image of an exoplanet, astronomers will probably have to resort to interferometry, a technique that combines the signals detected by widely spaced telescopes in order to reveal tiny details. NASA's Space Interferometry Mission, tentatively scheduled for 2009, should be able to detect exoplanets only a few times the size of the Earth. More sophisticated successors for exoplanet study may be launched in about 2015. NASA plans a mission called the Terrestrial Planet Finder and the European Space Agency plans one called Darwin. These will combine light collected by an array of mirrors using interferometry, not only to image Earth-sized planets but also to record good enough spectra to determine the compositions of their atmospheres.

An alternative strategy to document exoplanets would be to send autonomous probes to nearby stars. As presently envisaged, we could probably build and launch such a probe in about ten years. This would require maybe 20 years flight-time to reach a very nearby star, if, for example, accelerated to near light-speed by push from a laser beam. However, the probe would be travelling so fast when it got there that it would traverse any planetary system in only a few hours. Useful observations would thus be both brief and difficult to make, and would of course tell us about only one planetary system per probe. The rate of technological advance in signal processing seems to be such that over the coming 30 years it will be far more effective to study exoplanets using instruments in near-Earth orbit rather than dispatching interstellar probes. This is likely to change only in the improbable event that someone discovers a cheap and reliable interstellar propulsion system, preferably one that circumvents the theoretical limit to travel speed set by the speed of light.

Life on exoplanets

One of the goals of exoplanet studies is to determine whether any of them are capable of supporting life. Formerly it was assumed that life (or, at least, carbon-based life like our own) was likely to occur only on planets orbiting within the 'habitable zone' around a star. This is the region satisfying the so-called 'Goldilocks criterion', where the temperature is not too hot, not too cold, but just right. In the context of conventional life, 'just right' means that liquid water must be able to exist at the surface. In our own Solar System, only the Earth currently fits the bill, orbiting as it does near the middle of the Sun's habitable zone. Moreover, the Earth is massive enough to retain a decent atmosphere but not so massive as to have captured vast quantities of hydrogen and helium from the solar nebula and thus become a giant planet. The discovery of life in various extreme environments on Earth and the realization that life could be abundant below the icy surface of Europa suggests that terrestrial planets in a star's traditionally defined habitable zone are not the only feasible abodes of life. Even so these are probably where life is going to be most abundant and easiest to detect.

Given that our own galaxy contains about 100,000 million (10^{11}) stars, many people would accept that it would be very strange indeed if life were to occur only in our own Solar System. The number of potentially life-bearing planets in our own galaxy can be estimated as follows. The rate of star formation in our galaxy is known to be about 20 per year, and has probably been much the same for at least the past 5 to 10 billion years. About one star in ten is Sun-like. About half the Sun-like stars that have been studied are now known to have planets. This means that, in our own galaxy alone, about one planetary system is formed about a Sun-like star every year. To calculate the number of life-bearing planets that we would expect to exist in our galaxy today, this number must be multiplied by three factors:

- the average number of planets capable of bearing life in each planetary system
- the chances that life will begin on any planet capable of bearing life
- the length of time that life survives once established.

If we estimate the average number of planets capable of bearing life in each planetary system as one we will not be far out. The length of time that life survives once established (the third factor) is probably virtually the entire life of the planetary system: in our own Solar System life has survived (so far) for 4 billion years, so this is a reasonable value to use in our equation. The second factor, the chances that life will begin on any planet capable of bearing life, is the only one of the three open to considerable dispute. Some biologists claim that, given the right environment and a few tens of millions of years to play with, life will almost inevitably arise. If this is correct, then the number of life-bearing planets in our galaxy must be several billion. Other biologists say that even given favourable conditions, the odds against life starting by chance are exceedingly remote. However, these odds would need to be billions to one for life in our galaxy to occur only in our own Solar System, and tens of trillions to one for it to be completely absent in the tens of millions of other galaxies within the observable universe. The burgeoning field of study devoted to these topics, and also to investigating the survivability of terrestrial organisms in exotic environments, has become known as **astrobiology**.

This question of life-bearing planets in the nearby part of our own galaxy will be settled soonest by intruments such as the Terrestrial Planet Finder and Darwin, capable of making spectroscopic searches for oxygen, ozone and methane in the atmospheres of terrestrial exoplanets. These are highly reactive chemicals, that it is believed could not survive together. in any abundance without the action of a vigorous biosphere.

Is there anybody out there?

If life-bearing planets are indeed common, then the question arises of whether any intelligent life has arisen. In particular are there likely to be civilizations with whom we could communicate? More than forty years of searching the skies for signals from alien civilizations has failed to detect any, but the galaxy is a big place and it is far too early to conclude that we really are alone. Certainly, anything like a normal conversation by radio or other electromagnetic radiation would be impossible, because such a signal cannot travel faster than the speed of light, and so would take more than four years on a one-way journey even to the nearest star.

To estimate the number of technological civilizations currently in our galaxy, the above calculation has to be modified by replacing length of time that life survives by a smaller number. This is composed of the fraction of those planets on which life begins where intelligence evolves, multiplied by the fraction of those that develop technological civilizations wishing to communicate, multiplied by the average lifetime of a technological civilization. The expression modified in this way is known as the Drake Equation, after the American astronomer Frank Drake who framed it in 1961.

Of these three extra factors, the fraction of life-bearing planets where intelligence evolves is likely to be close to one, because intelligence confers a competitive (evolutionary) advantage to a species. The fraction of intelligence-bearing planets that develop a technological civilization that wishes to communicate is unknown, but a reasonable guess would be about half of them.

The average lifetime of a technological civilization is unfathomable. However, we have had a radio-using civilization of our own for over half a century and have so far managed to avoid blasting ourselves into oblivion by a nuclear war or destroying our society in a self-made environmental catastrophe. If the average lifetime of a technological civilization is taken to be 100 years, then the Drake equation suggests that there should be about 50 communicating civilizations currently in our galaxy, provided that life is virtually inevitable on planets capable of bearing life. However, the number could be as low as one (i.e. just us!). On the other hand, if we take a more optimistic view that technological civilizations last longer than a century, then the number of such civilizations in our galaxy increases proportionally.

One thing is sure though, irrespective of whether or not we really are alone. This is that there are plenty of planetary bodies in our own Solar System and beyond, each with its own unique landscapes, climate, history and (possibly) lifeforms. The study of planets is thus both fascinating and (unlike this book) never ending.

taking it further

If you have enjoyed this book, you may find two other books by myself in the *Teach Yourself* series of interest: *Teach Yourself Geology* discusses the full range of geological processes on planet Earth much more fully than in Chapter 05, and *Teach Yourself Volcanoes* explores volcanic processes on the Earth and other planetary bodies. By a different author, *Teach Yourself Astronomy* is a basic introduction to observing the sky and astronomy in general.

If you want a book that leads you further into study of the Solar System from the level reached in this one, the best single volume is *The New Solar System* (4th edition, 1999) edited by J. K. Beatty, C. C. Petersen and A. Chaikin and published jointly by Sky Publishing Corporation and Cambridge University Press. Fascinating personal accounts by many of the scientists involved in Solar System exploration during the 1980s and 1990s are gathered together in *Our Worlds* by S. A. Stern (1999, Cambridge University Press).

There are so many books about individual bodies such as the Moon, Mars and Venus, that I hesitate to draw attention to any one of them. Some are written by journalists and some by scientists. Each type has its virtues, and it is up to you to decide which suits you best. My favourite account of the glory days of lunar exploration culminating in the Apollo landings is *To a Rocky Moon: a Geologist's History of Lunar Exploration* by Don E. Wilhelms (1994, University of Arizona Press). Finally, if you have become intrigued by the large and diverse regular satellites of Jupiter and the other giant planets then you should consult my own *Satellites of the Outer Planets* (2nd edition, 1999) published by Oxford University Press.

Apart from reading about them, there are two supreme ways to find out more about the planets. One is to go outside and look at them for yourself. You might not see much detail, but the thrill is incomparable. The other is to take advantage of the vast amount of images and other information now freely available over the worldwide web.

Observing the planets

There are plenty of books about observing planets. Here there is space only for the merest of guidelines.

Firstly, there is the matter of locating planets in the sky. The Moon, of course, is obvious. In addition, if you are at all familiar with the patterns of stars in the night sky, you should have little difficulty in recognizing that Venus, Mars, Jupiter and Saturn are interlopers and at least as bright as the brightest stars. They are not always visible of course, because sometimes they lie more-or-less in the same direction as the Sun. As described in Chapter 03, it is quite a challenge to spot Mercury, because it never strays far from the Sun. If you have really no idea what planets ought to be visible, many newspapers carry a monthly astronomy column and star map describing the current night sky, and indicating the positions of the brighter planets. There are also web-based Solar System plotters that can give you similar information (see below).

To see one of the more distant planets or any of the asteroids requires optical aid, but very powerful instruments are required before any details are visible. However, if you have access to binoculars, these can reveal many features of the lunar landscape. Avoid times of full Moon, because no shadows are visible. Instead try about a week before or a week after full Moon, when shadows along the terminator (the sunrise or sunset line) are best seen from Earth. Binoculars work best if you can hold them steady. Preferably mount them on a tripod, but even steadying them by leaning against a wall will help. Steadily-held binoculars will also reveal the phases of Venus. Try this at twilight, rather than in a completely dark sky, to minimize the glare. They will also show Jupiter as something bigger than a point of light, and will reveal all four of its galilean satellites (though not necessarily all at the same time).

A telescope, because of its greater magnification, is useless for observing planets unless it is properly mounted. Something with a main lens or mirror just a few centimetres across and a magnification of ×30 will allow you to explore the Moon in detail, will show the flattened shape of Jupiter's disc and the rudiments of its atmospheric banding, and will reveal the rings of Saturn. Telescopes with apertures exceeding ten centimetres will show the polar caps on Mars and albedo patterns on its surface.

These days, many amateurs have access not just to telescopes but also to sophisticated telescope-mounted imaging devices. If this is an interest you would like to develop, then your best bet is to consult a monthly magazine such as *Sky and Telescope* or *Astronomy Now*.

Planets on the Web

The worldwide web (or Internet), to which increasing numbers of us now have access at work, at school or at home, is a tremendous resource of news and images for anyone interested in planets. Here are a few key website addresses. If you cannot find what you need at these sites, then links from them will almost certainly take you somewhere useful.

Solar system news, catalogues and simulations

Catalogues, news and information: **http://ssd.jpl.nasa.gov/**
Catalogue and images: **http://nssdc.gsfc.nasa.gov/planetary/**
Maps of solar system bodies and simulated views from a variety of viewpoints: **http://samadhi.jpl.nasa.gov/**
Small bodies (including asteroids): **http://pdssbn.astro.umd.edu/**
Minor planet center: **http://cfa-www.harvard.edu/iau/mpc.html**
Near-Earth asteroids: **http://earn.dlr.de/nea/**
Satellites of Jupiter and other giant planets: **http://jupiter.berkeley.edu**
Solar System exploration: **http://sse.jpl.nasa.gov/**
Asteroids in weird orbits: **http://www.astro.queenscu.ca/~wiegert/**

Galleries of planet images

General archives of NASA mission images: **http://photojournal.jpl.nasa.gov/**
http://wwwflag.wr.usgs.gov/USGSFlag/Space/wall/wall.html

http://pds.jpl.nasa.gov/planets/
The Nine Planets (multimedia tour of the Solar System):
http://seds.lpl.arizona.edu/billa/tnp/
Image-based maps of Mars, Venus, the Moon and various
other satellites:
http://pdsmaps.wr.usgs.gov
The Moon: http://cass.jsc.nasa.gov/moon.html
Comet Shoemaker Levy 9 impacts with Jupiter:
http://www.jpl.nasa.gov:80/sl9/

Current and past missions to planets

General: http://sse.jpl.nasa.gov/missions
Mars exploration: http://mpfwww.jpl.nasa.gov/
Clementine (Moon):
http://www.nrl.navy.mil/clementine/clementine.html
Magellan (Venus):
http://www.jpl.nasa.gov/magellan/mgn.html
Near Earth Asteroid Rendezvous (Eros and Mathilde):
http://sd-www.jhuapl.edu/NEAR/
Voyager missions (Jupiter–Neptune): http://voyager.jpl.nasa.gov/
The Jupiter system as explored by Galileo:
http://galileo.jpl.nasa.gov/
The Cassini mission to Saturn:
http://saturn.jpl.nasa.gov/index.cfm
http://ciclops.lpl.arizona.edu/
The Huygen's Titan entry probe: http://sci.esa.int/huygens/
The Beagle 2 Mars lander:
http://beagle2.open.ac.uk/index.htm
Mars Express: http://sci.esa.int/marsexpress/

Future missions to planets

Nozomi (Japanese Mars mission):
http://www.isas.ac.jp/e/enterp/missions/nozomi/cont.html
American mission to Mercury: http://messenger.jhuapl.edu/
European mission to Mercury:
http://sci.esa.int/home/bepicolombo/index.cfm
Jupiter Icy Moons Orbiter: http://www.jpl.nasa.gov/jimo/
New Horizons Pluto-Kuiper belt mission:
http://pluto.jhuapl.edu
Dawn (to Ceres and Vesta):
http://www.ssc.igpp.ucla.edu/dawn/

Lunar A (Japanese lunar mission):
http://www.isas.ac.jp/e/enterp/missions/lunar-a/index.html
Mars exploration (general NASA site):
http://mars.jpl.nasa.gov/
Mars Exploration Rovers (NASA):
http://mars.jpl.nasa.gov/mer/
Muses-C (Japanese asteroid sample return):
http://www.isas.ac.jp/e/enterp/missions/muses-c/index.html
Nozomi (Japanese Mars orbiter):
http://www.isas.ac.jp/e/enterp/missions/nozomi/index.html
Rosetta (ESA comet rendezvous): http://sci.esa.int/rosetta/
Selene (Japanese lunar mission):
http://www.isas.ac.jp/e/enterp/missions/selene/index.html
SMART 1 (ESA lunar orbiter): http://sci.esa.int/smart/
Venus Express (ESA mission): http://sci.esa.int/venusexpress/

Space Telescope images

http://www.stsci.edu/resources

Asteroids and comets

http://www.hohmanntransfer.com/news.htm
http://www.astro.uwo.ca/~wiegert/

The Kuiper belt

http://www.boulder.swri.edu/ekonews
http://www.ifa.hawaii.edu/faculty/jewitt/kb.html

Extrasolar planets

http://www.obspm.fr/encycl/encycl.html

Astrobiology

News and information: http://www.astrobiology.com

Search for Extraterrestrial Intelligence

SETI Institute: http://www.seti-inst.edu/
SETI@home experiment: http://setiathome.ssl.berkeley.edu

Space agencies

NASA: http://www.nasa.gov
ESA: http://www.esa.int/ and http://www.esrin.esa.it
ISAS (Japan): http://www.isas.ac.jp
IKI (Russia): http://arc.iki.rssi.ru/eng/index.htm

Societies and organizations

International Astronomical Union: http://www.iau.org

Division of Planetary Sciences (American Astronomical Society): http://www.aas.org/~dps/dps.html

Lunar & Planetary Institute: http://cass.jsc.nasa.gov/lpi.html

Planetary Forum: http://ast.star.rl.ac.uk/forum/

British Astronomical Association (a mainly amateur body): http://www.britastro.org

Royal Astronomical Society (a mainly professional body): http://www.ras.org.uk

The Planetary Society: http://www.planetary.org

albedo The fraction of sunlight that falls on a body which is reflected by it.

aphelion The point on an orbit around the Sun that is furthest from it.

asteroids Rocky or metal-rich bodies of irregular shape, which range in size from a few hundred km in diameter downwards. Most occur between the orbits of Mars and Jupiter.

asthenosphere The weak-and-mobile interior part of a planetary body's mantle, which, although solid, is sufficiently warm and mobile to convect slowly.

astrobiology A wide field of study, ranging from the origin of the building blocks of life to theoretical and practical research into extraterrestrial habitats for life.

Astronomical Unit (AU) The average distance between the Earth and the Sun (150 million km), which is a useful unit for comparing distances within the Solar System.

axial inclination The amount by which a planet's rotation axis departs from being perpendicular (at right angles) to its orbital plane. If less than 90° the rotation is prograde, if greater than 90° it is retrograde.

basalt A variety of lava that is typical of the Earth's ocean floors, the lunar maria, much of the surface of Venus, and the volcanic regions of Mars. It is poorer in silica than most other varieties of lava, and this gives it a comparatively low viscosity, so that it flows freely and gases can escape gently rather than triggering an explosive eruption.

carbonaceous chondrite A kind of rocky meteorite, containing hydrated (water-bearing) minerals and up to several per cent by weight of carbon in the form of (non-biological) organic molecules. These are thought to be samples of the most 'primitive', least-altered, material from which the Solar System grew.

comet An object typically up to a few tens of km across composed of a mixture of carbonaecous and icy material, in a highly eccentric (elongated) orbit about the Sun. When a comet's orbit brings it close to the Sun some of its ice is evaporated, and the comet may develop a spectacular 'tail' of gas and dust pointing away from the Sun that can become up to 100 million km long in extreme cases.

corona (plural: coronae) On Venus circular or elliptical features, typically a few hundred km in diameter marked by a pattern of concentric fractures and with volcanic activity concentrated near their centres. Thought to have developed above zones of upwelling within the mantle. On Miranda, a large concentrically or complexly patterned terrain unit of uncertain origin, but apparently unrelated to the coronae of Venus.

core The dense inner region of a planetary body that has become compositionally differentiated, and which lies below the mantle. The cores of the terrestrial planets are iron-rich. Large icy satellites and giant planets have rocky cores, which may have iron-rich inner cores within them.

cratering timescale The relationship between the number of impact craters per unit area of planetary surface and the age of that surface. The timescale has been calibrated in absolute terms using samples of independently known age from the Moon, and is applied with reasonable confidence throughout the inner Solar System. However, among the asteroids and satellites of the outer planets different populations of impactors have contributed to crater formation, so absolute ages cannot be determined from crater-counting alone. However, the principle that, on a given planetary body, older surfaces have more craters per unit area than younger surfaces still applies.

crust The outermost, chemically distinct, layer in a solid planetary body. In a terrestrial planet the crust is rock, and is richer in low-density elements than the underlying mantle. In an icy satellite or similar body, the crust (where present) is richer in salts and volatile ices than the underlying icy mantle.

cryovolcanism Volcanism involving melts derived from various kinds of ice (*kryos* is the Greek word for frost).

differentiation The segregation of the interior of a planetary body into layers of different composition. Typically, denser materials sink towards the centre (to form a core) and less-dense materials rise upwards to form the mantle that surrounds the core. The crust is the outermost layer of a solid planetary body that can sometimes be recognized as chemically differentiated from the mantle.

eccentricity A measure of the degree to which an ellipse is elongated, and therefore departs in shape from a circle (which is an ellipse with an eccentricity of zero). Technically defined as the ratio between the distance between the two foci of the ellipse and the length of its long axis.

ecliptic The plane of the Earth's orbit about the Sun, close to which lie the orbits of most other bodies in the Solar System.

ejecta Generally fragmented material thrown out of an impact crater by shock waves during its formation. Ejecta may form a recognizable blanket-like deposit close to a crater, but further away is limited to isolated blocks. Some large blocks of ejecta may be big enough to cause secondary craters of their own when they hit the ground.

elongation The apparent separation between a planetary body and the Sun, as seen in the sky. Mercury and Venus, whose orbits lie inside the Earth's, are best seen when at their maximum elongation from the Sun.

exoplanet A planet orbiting a star other than the Sun, otherwise known as an extrasolar planet.

gas giants The two largest giant planets (Jupiter and Saturn) that have much deeper gaseous outer layers than the other two giant planets.

giant impacts Collisions between similarly sized objects, especially planetary embryos during the birth of the Solar System.

giant planets The four large planets in the outer Solar System (Jupiter, Saturn, Uranus and Neptune), that are tens or hundreds of Earth-masses and have no solid surface. Also a useful term for planets of another star (expoplanets) that are similar in mass.

gravity assist trajectory The path of a spacecraft chosen so that it passes sufficiently close to a planetary body for its course and speed to be changed as it swings through the body's gravitational field.

greenhouse effect   Warming of a planetary surface by virtue of the atmospheric abundance of 'greenhouse gases'. These are molecules containing more than one kind of atom (such as carbon dioxide and water vapour), which makes them efficient at absorbing infrared radiation.

Hadley cell   Part of the pattern of atmospheric circulation. Warmed air expands and rises, flows polewards, cools, contracts and sinks to return equatorwards at low altitude. On Venus there is a single Hadley cell in each hemisphere, extending from the equator to the poles, whereas Earth's more rapid rotation breaks the pattern into a series of three Hadley cells between the equator and each pole.

IAU   International Astronomical Union, the internationally recognized authority for assigning designations to celestial bodies and their surface features.

ice   Sometimes used simply to refer to frozen water, this can also mean other volatiles in a frozen state, such as methane, ammonia, carbon monoxide, carbon dioxide, and nitrogen (either individually, or mixed together).

igneous   Referring to rock that has solidified from a molten state.

impact crater   A usually circular depression with a raised rim excavated by impact on a planetary surface by an asteroidal or cometary fragment, arriving at a speed of a few tens of km per second. The shock of the impact excavates the crater and flings out fragmented ejecta. The diameter of the crater is thought to be about 30 times the size of the impacting body.

Kepler's laws of planetary motion   First law: planets move in elliptical orbits, with the Sun at one focus of the ellipse. Second law: for a given planet, a line drawn from the planet to the Sun will always sweep out the same amount of area over a given time interval. Third law: the square of a planet's period of revolution round the Sun (its orbital period) is proportional to the cube of its average distance from the Sun.

Kuiper belt   A region extending outward from about the orbit of Neptune, concentrated at about 30–50 AU from the Sun, where there are large numbers of icy bodies of sub-planetary size known (with the exception of Pluto and its satellite Charon, which are the largest) as Kuiper belt objects. The Kuiper belt was independently predicted on theoretical grounds by Kenneth

Edgeworth (1943) and Gerard Kuiper (1951) and is alternatively known as the Edgeworth-Kuiper belt (or disc). The objects within it are referred to as Kuiper belt objects (KBOs) or Edgeworth-Kuiper objects (EKOs) or Trans-Neptunian objects (TNOs).

lava Molten material that wells up from below and flows across the surface of a planetary body before solidifying. The term lava flow can be applied either to a flow of lava while it is still moving, or to the same feature after it has completely solidified but remains identifiable as a distinct unit.

lithosphere The strong and rigid outer shell of a solid planetary body, comprising the crust (if any) and the outermost mantle. Below the lithosphere the mantle is capable of convection, and is sometimes distinguished by the term asthenosphere.

lunar highlands The oldest exposed regions of the Moon's crust, 4.5 to 3.8 billion years old. The highlands are composed of the rock type known as anorthosite, and are paler in appearance than the maria.

mantle The compositionally distinct material lying outside the core of a solid planetary body. Terrestrial planets have rocky mantles, icy satellites have icy mantles. In some cases, the outermost solid material is chemically slightly different from the bulk of the mantle and is referred to as the crust.

mantle plume A pipe-like upwelling within the asthenospheric part of the mantle, which when it reaches the base of the lithosphere heats it, and therefore causes it to expand. Volcanism results mainly from the escape of magma from within the plume itself but is also produced by partial melting of the lithosphere heated by the plume. A site situated above a mantle plume is sometimes described as a 'hot spot'.

mare (plural: **maria**) Low lying parts of the Moon that were flooded by basalt lava (mostly between 3.8 and 3.1 billion years ago), giving them a darker appearance than the lunar highlands.

metamorphic Referring to rock formed by recrystallization (but without melting) of a pre-existing rock because of the effects of pressure and/or heat.

meteorite A lump of rock or rock plus iron, or nickel-iron alloy that falls to Earth. Most are fragments from the asteroid belt, but a few have been identified that were blasted off the surfaces of the Moon and Mars during the formation of impact craters.

mineral A naturally occurring crystalline substance of a limited range of composition that, usually in combination with minerals of other compositions, makes up rock. The most common minerals in the terrestrial planets are silicates, which consist of silicon and oxygen, usually combined with various metallic elements.

minor planet Formerly just a synonym for asteroid, this term now has special value because it can describe any small body orbiting the sun and so also includes Kuiper Belt objects and bodies of debatable affiliation, such as Centaurs.

multiringed impact basin A structure of concentric fractures produced by the impact of an asteroid or comet that is too large to form a conventional crater. Typical diameters range from several hundred to a couple of thousand km.

NASA National Aeronautics and Space Administration, the United States civilian agency responsible for space exploration.

occultation The passage of a dim object in front of a brighter one, so that the light of the distant object is dimmed. Examples include a planetary ring of a distant star or an exoplanet in front of its own star.

opposition The situation when a planetary body lies in the opposite direction to the Sun, as seen from Earth.

orbital resonance A state in which one orbiting body is subject to periodic gravitational perturbations by another, because the orbital period of one is a simple ratio (e.g. 2:1, 3:2, 4:1 etc) of the other's.

peridotite A rock type having the same composition as the Earth's mantle. Silica content is below approximately 45%, which is less than that in basalt.

perihelion The point on an orbit around the Sun that is closest to it.

photodissociation The break-up of otherwise stable atmospheric molecules of gas under the influence of solar ultraviolet radiation. Once a molecule has been split in this way, the components are usually highly reactive and available to take part in chemical reactions (a very important process in controlling atmospheric composition). Light components, such as hydrogen, may be able to escape from the atmosphere altogether.

planet   One of the large objects orbiting the Sun (or any other star). Nine bodies (Mercury, Venus, Earth, Mars, Jupiter, Saturn, Uranus, Neptune and Pluto) are conventionally classed as planets in our Solar System. A precise definition is not helpful because Pluto appears just to be an exceptionally large Kuiper belt object (most of which are too small to be counted as planets) whereas several satellites of the planets have sizes, compositions and other characteristics that make them planets in all but name.

planetary body   A useful term encompassing planets, their satellites, asteroids, and other minor planets.

planetary embryos   During the birth of the Solar System, objects that formed by amalgamation of collided planetesimals, a few thousand km in size.

planetesimals   Objects that grew within the solar nebula, tens to hundreds of km in size.

plate tectonics   A description of how the rigid plates into which the Earth's lithosphere is divided are created, migrate, and are destroyed.

precession   The slow change in the direction to which a planet's rotational axis points.

prograde   Orbital or rotational motion that is anticlockwise when viewed from above the Sun's (or the Earth's) north pole. In the case of a satellite, orbital motion is defined as prograde if it is in the same direction as its planet's spin. Most motion in the Solar System is prograde.

radiogenic heating   Heating of a planetary body caused by the decay of radioactive isotopes. Abundant elements that have radioactive isotopes whose decay generates a significant amount of heat are uranium, thorium and potassium.

radio occultation   A way of investigating the properties of an atmosphere by studying the fading of the radio signal from a spacecraft as it passes behind a planetary body as seen from Earth.

regolith   Surface material or 'soil' that has been fragmented and widely distributed as a result of impact cratering.

retrograde   Orbital or rotational motion that is clockwise when viewed from above the Sun's (or the Earth's) north pole, and is thus in the opposite sense to prograde motion. In the case of a satellite, orbital motion is defined as retrograde if it is in the opposite direction to its planet's spin.

Roche limit   The critical orbital distance from a planet within which a hypothetical satellite composed of very weak material would be pulled apart by tidal forces. The same limit applies more or less to real large satellites regardless of their actual strength, but small satellites can survive slightly within a planet's Roche limit.

rotation   The spin of a planetary body about an axis running through its pole.

satellite   A body in orbit around another. Technically speaking the planets are satellites of the Sun, but the term is usually reserved for bodies (natural or artificial) in orbit about another, larger, planetary body. There are several large satellites in the Solar System (our own Moon among them) that share all the important attributes of planets.

sedimentary   Referring to rock that has formed from an accumulation of fragments of pre-existing rock that have been transported from elsewhere.

shield volcano   A volcano with gently sloping flanks, formed mainly of basaltic lava.

silicate   A mineral composed of the elements silicon and oxygen, usually in combination with various metallic elements. The term also describes rock made out of silicate minerals, or magma within which silicate minerals would crystallize upon cooling.

solar nebula   The cloud of gas and dust from which the Sun and the rest of the Solar System formed.

Solar System   The Sun and the bodies gravitationally bound to it (i.e. the planets, their satellites, asteroids, Kuiper belt objects, comets etc).

spectroscopy   A means of establishing the presence and abundance of a substance by means of characteristic absorption lines in the spectrum of ultraviolet, visible or infrared light reflected (or emitted) by a surface or atmosphere.

synchronous rotation   A tidally induced situation, in which a satellite's rotation period is exactly the same as its orbital period. It therefore always presents the same face towards its planet.

terrestrial planets   The Earth and bodies like it (iron-rich core and rocky exterior); namely the planets Mercury, Venus and Mars, but also (on a broader definition) the Moon and Jupiter's large rocky satellite Io.

tholins   A large class of tarry substances produced from carbonaceous material, methane ice, or hydrocarbon-rich atmospheres under the influence of charged particles or ultraviolet radiation. Apparently common on the surfaces of small bodies in the outer Solar System and probably on Titan too.

tidal heating   Heating of a planetary body resulting from its shape being continually deformed by tidal forces. This process has had dramatic consequences for many of the satellites of the giant planets, particularly those in close orbit.

Trans-Neptunian object (TNO)   A planetary body belonging to the Solar System but orbiting beyond the orbit of Neptune. This includes classical Kuiper belt objects, Plutinos and Scattered Disk objects.

Trojan points   Also known as Lagrange points L4 and L5, these are two points on the orbit of a planet orbiting a star, or of a satellite orbiting a planet, that are 60 degrees ahead and 60 degrees behind the main orbiting body. Several much smaller bodies can share the orbit of the larger body provided that their mean position is very close to either of these Trojan points.

index

teach® yourself

Hinduism
History, 101 Key Ideas
How to Win at Horse Racing
How to Win at Poker
HTML Publishing on the WWW
Human Anatomy & Physiology
Hungarian
Icelandic
Indian Head Massage
Indonesian
Information Technology, 101 Key Ideas
Internet, The
Irish
Islam
Italian
Italian, Beginner's
Italian Grammar
Italian Grammar, Quick Fix
Italian, Instant
Italian, Improve your
Italian Language, Life & Culture
Italian Verbs
Italian Vocabulary
Japanese
Japanese, Beginner's
Japanese, Instant
Japanese Language, Life & Culture
Japanese Script, Beginner's
Java
Jewellery Making
Judaism
Korean
Latin
Latin American Spanish
Latin, Beginner's
Latin Dictionary
Latin Grammar
Letter Writing Skills
Linguistics
Linguistics, 101 Key Ideas
Literature, 101 Key Ideas
Mahjong
Managing Stress
Marketing
Massage
Mathematics
Mathematics, Basic
Media Studies
Meditation
Mosaics
Music Theory
Needlecraft
Negotiating
Nepali

Norwegian
Origami
Panjabi
Persian, Modern
Philosophy
Philosophy of Mind
Philosophy of Religion
Philosophy of Science
Philosophy, 101 Key Ideas
Photography
Photoshop
Physics
Piano
Planets
Planning Your Wedding
Polish
Politics
Portuguese
Portuguese, Beginner's
Portuguese Grammar
Portuguese, Instant
Portuguese Language, Life & Culture
Postmodernism
Pottery
Powerpoint 2002
Presenting for Professionals
Project Management
Psychology
Psychology, 101 Key Ideas
Psychology, Applied
Quark Xpress
Quilting
Recruitment
Reflexology
Reiki
Relaxation
Retaining Staff
Romanian
Russian
Russian, Beginner's
Russian Grammar
Russian, Instant
Russian Language, Life & Culture
Russian Script, Beginner's
Sanskrit
Screenwriting
Serbian
Setting up a Small Business
Shorthand, Pitman 2000
Sikhism
Spanish
Spanish, Beginner's
Spanish Grammar
Spanish Grammar, Quick Fix

Spanish, Instant
Spanish, Improve your
Spanish Language, Life & Culture
Spanish Starter Kit
Spanish Verbs
Spanish Vocabulary
Speaking on Special Occasions
Speed Reading
Statistical Research
Statistics
Swahili
Swahili Dictionary
Swedish
Tagalog
Tai Chi
Tantric Sex
Teaching English as a Foreign Language
Teaching English One to One
Teams and Team-Working
Thai
Time Management
Tracing your Family History
Travel Writing
Trigonometry
Turkish
Turkish, Beginner's
Typing
Ukrainian
Urdu
Urdu Script, Beginner's
Vietnamese
Volcanoes
Watercolour Painting
Weight Control through Diet and
   Exercise
Welsh
Welsh Dictionary
Welsh Language, Life & Culture
Wills and Probate
Wine Tasting
Winning at Job Interviews
Word 2002
World Faiths
Writing a Novel
Writing for Children
Writing Poetry
Xhosa
Yoga
Zen
Zulu

available from bookshops and on-line retailers

teach
yourself

volcanoes
david rothery

- Do you want to discover the mysteries behind volcanic activity?
- Do you want to know why volcanoes occur and how they erupt?
- Are you looking for a practical and comprehensive guide?

Volcanoes describes the processes involved in volcanic eruptions
and building volcanoes. It discusses the interactions between
volcanoes and the environment and climate and the hazards posed
by volcanoes. It also presents the various kinds of volcanoes on the
Earth and on other planetary bodies, and shows how volcanic
activity can be monitored and predicted.

David Rothery is a volcanologist at the Open University. His
international collaborative research on active volcanoes has taken
him to places as diverse as Hawaii, the Americas, Indonesia
and Italy.

teach
yourself

astronomy
patrick moore

- Are you interested in astronomy but don't know where to start?
- Are you an amateur astronomer wanting to play a role in this important science?
- Do you want to know the basic facts?

Astronomy is a practical answer to all your astronomy questions. It will help you to discover how to observe the Sun, Moon, planets and other bodies and introduce you to all that you can see with the naked eye and inexpensive equipment. Extensively illustrated throughout, the book also contains a selection of photographs from Patrick Moore's private collection.

Patrick Moore has a private observatory at Selsey, Sussex, and since 1957 has presented *The Sky at Night* on BBC TV every four weeks.

teach yourself

human anatomy & physiology
david le vay

- Do you need to know basic anatomy for a course or profession?
- Are you interested in how the body works?
- Do you want to understand more about scientific innovation?

Human Anatomy and Physiology is a comprehensive introduction to the structure and function of the human body. Extensively illustrated, the book also covers modern methods of investigation, relevant aspects of modern genetics, sports injuries, environmental and evolutionary considerations and the physiological aspects of AIDS.

David le Vay MS, FRCS was a consultant surgeon for many years and a well-known medical author and editor.